CRITICA

THE WORKS OF JAMES RADA, JR.

Battlefield Angels

"Rada describes women religious who selflessly performed life-saving work in often miserable conditions and thereby gained the admiration and respect of countless contemporaries. In so doing, Rada offers an appealing narrative and an entry point into the wealth of sources kept by the sisters."

Catholic News Service

Between Rail and River

"The book is an enjoyable, clean family read, with characters young and old for a broad-based appeal to both teens and adults. *Between Rail and River* also provides a unique, regional appeal, as it teaches about a particular group of people, ordinary working 'canawlers' in a story that goes beyond the usual coverage of life during the Civil War."

Historical Fiction Review

Canawlers

"A powerful, thoughtful and fascinating historical novel, *Canawlers* documents author James Rada, Jr. as a writer of considerable and deftly expressed storytelling talent."

Midwest Book Review

"James Rada, of Cumberland, has written a historical novel for high-schoolers and adults, which relates the adventures, hardships and ultimate tragedy of a family of boaters on the C&O Canal. ... The tale moves quickly and should hold the attention of readers looking for an imaginative adventure set on the canal at a critical time in history."

Along the Towpath

OTHER BOOKS BY JAMES RADA, JR.

Fiction

Beast

Between Rail and River

Canawlers

Kachina

Kuskurza

Logan's Fire

My Little Angel

The Race

The Rain Man

October Mourning

Non-Fiction

Battlefield Angels: The Daughters of Charity Work
as Civil War Nurses

Kidnapping the Generals: The South's Most-Daring Raid
Against the Union

Looking Back: True Stories of Mountain Maryland

Looking Back II: More True Stories of Mountain Maryland

When the Babe Came to Town: Stories of George Herman
Ruth's Small-Town Baseball Games

SAVING SHALLMAR

Christmas Spirit in a Coal Town

by
James Rada, Jr.

Ryan,
Santa lives in
Shallmar

LEGACY
PUBLISHING

A division of AIM Publishing Group

SAVING SHALLMAR:
CHRISTMAS SPIRIT IN A COAL TOWN

Published by Legacy Publishing, a division of AIM Publishing
Group.
Gettysburg, Pennsylvania.
Copyright © 2012 by James Rada, Jr.
Printed in the United States of America.
First printing: November 2012.

ISBN 0-9714599-7-5

Cover design by Stephanie E. J. Long.

LEGACY
PUBLISHING

315 Oak Lane • Gettysburg, Pennsylvania 17325

For my great mom, Sylvia Gale Rada,
my biggest fan and my staunchest supporter.
Love you!

1

A True Story of Santa Claus

Everyone who lived in the tiny coal-mining town of Shallmar, Md., in 1949 believed in Christmas from nine-month-old Walter Hedrick, born about the time the Wolf Den Coal Corporation shut down in March, to fifty-nine year old Howard Marshall, president of the Wolf Den Mining Corporation and probably the oldest man in town. Every last one of them knew that the Spirit of Christmas was a real thing even if he or she had no use for Mary, Joseph and Jesus.

They even believed in Santa Claus.

They would tell you that Santa, well; he really wasn't a jolly, old, fat man with a flowing white beard who dressed in red velvet with white fur trim. And he didn't zip around the world in one night in a sleigh pulled by eight flying reindeer (nine, if you counted the red-nosed buck that Gene Autry started singing about in November).

No way was that Santa!

The people of Shallmar were coal miners who lived hard lives marked by bruises and broken bones that were all too often cut short by a cave in or other mining accident. So they had no time for such foolishness and children's tales.

They would tell you that Santa Claus was a rather plain-looking man of average height. He had thinning brown hair, wide shoulders, no beard and he wore glasses. Nor was he fat and his suits were the type worn with a necktie. Though he might have been able to use a sleigh during the winters in the Maryland mountains, he found it more practical to drive the beat-up 1937 Plymouth coupe his wife had bought new before they were married. It was one of the few cars in town.

You see, Santa Claus lived in Shallmar.

The people of Shallmar knew this for a fact because they saw him work his Christmas magic in 1949, the year the town was saved from dying.

You might wonder if a town can really die. Well, if the people in the town die off or simply leave when they have the ability to do so,

1

then no one is left in town. Such a place is called a ghost town, and since something that's living can't be a ghost, a ghost town is a town that has died. The homes have no heartbeats.

That was Shallmar. Or rather, that's what Shallmar was speeding towards in 1949 faster than ol' St. Nick supposedly circled the globe on Christmas Eve night. Nobody realized it at first, or maybe it was that nobody wanted to admit what was happening, much less do anything about it.

Except for one man. His family called him June. His friends called him Junior. His wife called him J. Paul or just J. P. His students called him Mr. Andrick. And while those were all names he answered to, he was best known as Santa Claus.

Not that anyone called him that. Most people just called him Paul.

J. Paul Andrick

2

Towns Come and Go

But the Mountain Remains

Slaves, criminals and serfs mined coal until the fifteenth or sixteenth centuries, but they didn't do it by choice. Coal mining wasn't considered a job or trade back then; it was a punishment. Some would say that it never stopped being that; that miners are still essentially serfs and slaves.

Country singer Merle Travis sung about the coal mining life in his 1946 song, "Sixteen Tons." The chorus of the song captured the way that coal miners felt trapped into risking their lives daily whether they wanted to or not:

"You load sixteen tons an' whaddya get?
Another day older an' deeper in debt.
Saint Peter doncha call me 'cause I can't go.
I owe my soul to the company sto'."

Coal didn't become commonly used in America until the mid-1800s. The earliest reference to coal and a coal mine in Western Maryland that's been found so far is on a 1751 Frye and Jefferson map of Virginia. On a spot near George's Creek in present-day Allegany County are the words "Coal Mine."

These first coal deposits were found close to the surface. That only made sense, seeing as how no one had much use for coal at the time so they weren't going to waste time looking for it. They had more-pressing things on their minds like getting crops planted and harvested. It was only when the spring rains and floods washed away the topsoil exposing the soft coal that it caught the attention of farmers in Western Maryland.

Farmers mined it part-time in the early 1800s and then only because they had nothing else to do in the winter. Their operations were small because they only needed enough coal to heat their homes. They couldn't sell it, not with the people most likely to buy coal being days away on the National Road. So why dig up more than was needed?

That all changed with the arrival of the Baltimore and Ohio Railroad in Cumberland, Maryland, on November 5, 1842. People came from far and wide to see the arrival of the first train into Cumberland. Businesses closed up early so their employees could be at the big event. People felt something big was coming and they wanted to celebrate.

And why not? The train promised to connect their city to Baltimore in the east and since Baltimore was a port city, the world was suddenly within reach. More importantly, it could be reached within a reasonable time. For merchants and farmers, it meant their goods could reach markets quicker. Markets that had been too expensive to reach by hauling goods over the National Road would be affordable.

The crowd gathered near where the railroad tracks crossed Baltimore Street in the center of Cumberland. Bands played. Officials gave speeches. It was *the* event of the year.

Then the first train appeared, blowing smoke and cinder, and wheels squealing against the iron rails. The sight and sound of it all not only scared nearby horses, but many of the people gathered for the celebration screamed in terror. Very few of them had ever seen anything like it before that time.

When the train rolled to a stop in Cumberland shortly before 5 p.m., it had completed the normally days-long journey from Baltimore in ten hours.

Cumberland was just the beginning. The B&O Railroad continued its move west connecting cities hundreds of miles apart with quick and inexpensive transportation. From Cumberland, the railroad passed through Garrett County. Small coal communities sprang up near the rail line in the county beginning in 1876, right around the time the U.S. was entering its industrial age and all of those manufacturing factories needed coal – lots of coal – to power their furnaces.

Coal is burned in furnaces attached to boilers. The heat from the burning coal turns the water in the boilers to steam, which then turns turbines to run ships and machinery. Coal is used in smelting furnaces to create fires hot enough to melt iron so that it can be refined into steel. For both these reasons, a lot of coal was needed as America grew in the late 1800s and coal mining quickly grew into a major industry in Garrett County.

The Appalachian Mountains on the western end of Maryland are where the state's coal industry resided and still does, though on a much smaller scale. Western Maryland is part of the Appalachian coal field, which also includes West Virginia, Pennsylvania, eastern Kentucky, western Virginia, central Tennessee and northern Alabama. It generally produces high-quality bituminous coal that is used in furnaces to generate power. The other two coal-producing areas in the United States are the Midwestern and Rocky Mountain coal fields. Western Kentucky, Indiana and Illinois make up the Midwestern coal field and Wyoming, Montana and New Mexico make up the Rocky Mountain coal field. Midwestern coal has high sulfur content and is burned to make electricity. Rocky Mountain coal is the lowest-grade coal in the country, but its low-sulfur content allows it to be cleaner burning.

The Maryland coal industry wasn't centered in one central headquarters complex in Garrett and Allegany counties, but in the small mining patches and towns along the border of Allegany and Garrett counties and along the North Branch Potomac River.

They had names like Vindex, Dodson, Kempton and Kitzmiller. Some of these towns lived only a few short years, dying when the coal in the mine played out just as the flames that consumed the coal would do. Other towns found reasons besides coal mining to live and so they continued on even after the coal was gone.

When one coal camp died, another town would soon take its place wherever the next coal seam was uncovered. The miners tended to follow the coal because that is where they could find work. As long as there was a demand for coal, there would be those who sought where it hid. Once finding it, a coal company was established to shoot, cut or dig that coal from the mountains.

As the company grew, houses were built for the miners to live in because towns were rarely already in existence where coal was found. Coal patches tended to be in remote areas, not because of the coal dust and noise associated with a mining operation, but because that was where the coal seams led. The mining company would also build a store to provide miners with food, clothes, tools and other things. From there, the coal mine sites would get churches, schools, other businesses and eventually a town was created.

Coal companies used the isolation of the coal towns to their benefit. Since the companies owned everything in the towns, the company had a power – a control – over the miners. Mine superintendents could fire and evict troublemakers and ensure they couldn't find another job in the coal town. That's why I said miners were still serfs even if they weren't called that.

As one Kempton miner named Harrison Hanlin put it, "Everyone had to live accordingly." And the coal companies determined what was "accordingly."

Miners waiting to get their mail at the Kempton company store. Company stores like this one served as a post office and general store. Shallmar's company store was the only place many miners in Shallmar could get to to buy their food and clothing. Courtesy of the Garrett County Historical Society.

Shallmar was just one of these many small coal towns near the railroad line. The Wolf Den Coal Company purchased the land from the Manor Mining Company, which hadn't managed to develop a mine on the land. It started operation in July 1917, but the news of a new company opening in Garrett County was lost amid news of stories of the country preparing for war and the military draft. The company made its first coal shipment in April 1918.

It wasn't the best time to start a coal-mining company, though you would have been hard pressed to tell it. The country had just gotten into World War I – The Great War – and coal was needed to power the massive steamships that carried American troops overseas and the warships that would defend freedom. In general, the war was a boom time for coal companies with plenty of demand for coal.

That was the problem with coal. Though it was an essential commodity at the time, demand for coal remained about the same unless something happened to drive it up like a war. War not only increased demand, but it reduced the supply of coal available because many diggers became soldiers fighting in the war. The Wolf Den Coal Company even had to hire women during World War I to try and keep up with demand for coal. The women didn't actually dig coal in the mine (that wouldn't actually happen in U.S. coal mines until 1973), but they did work on the picking table sorting the coal from the rock. Using women was virtually unheard of and the male miners teased the women miners of Shallmar a lot, or worse, some of the older miners considered it bad luck for a woman to be in a coal mine. It got so bad that the male miners had to start finding alternative routes to the mine because the women would wait along the path and bombard the men with rocks in retaliation for the teasing.

With the increased demand for coal in World War I, the peak production year in the United States for bituminous coal was in 1918, though the Wolf Den Mine's peak year wasn't until 1929. Still, the writing was on the wall. When the war ended, so did the heavy demand. By 1923, U.S. coal mines were only producing half of what they were capable of producing. That had an effect on a miner's pay because they were working less. Miners had worked an average of 249 days a year during the war. By 1921, they worked an average of 142 days that year.

Shallmar may have started out small and grown smaller, but among coal mining towns, it was considered *the* place to live at one time. I'm not talking about just among the coal camps of Western Maryland but among the coal towns in Pennsylvania and West Virginia, too. You might call it coal country's Garden of Eden. The difference was that while Adam and Eve were cast out of the Garden of Eden, the people of Shallmar stayed. It was the town that slipped away from them, almost unnoticed.

Despite the fact that its name made it sound like a French country village, it's doubtful that any Frenchman ever lived in Shallmar. Garrett County, particularly the coal region, was made up of a lot of Scotch, Italians and Poles with a smattering of Germans and Mexicans thrown in. The town's population was never more than 500 people with only 100 of them being mining company employees. The rest were wives and children.

Shallmar in the winter seen from atop the surrounding mountains. The town of Shallmar got its name from switching the syllables of the last name of the town's founder, Wilbur Marshall. Photo courtesy of Jerry Andrick.

Shallmar was different from other coal patches in that it hadn't grown haphazardly around a coal mine. When Wilbur Marshall decided to start large-scale mining operations along the North Branch Potomac River, he laid out a town as well as started a coal company.

At forty years old, Wilbur was already a successful businessman in New York City when he started the Wolf Den Coal Company. Besides being president of Wolf Den Coal Company, he was president of W. A. Marshall and Company, Lincoln Coal Company and Bethel Realty Company. He was also the treasurer for the Lincoln Realty Company. Not too bad for a man who trained to become a carpenter as a young man. He moved from carpentry to the coal business in 1893 when he lived in Boston. Since then, he had been building on his successes to create additional companies.

Wilbur Marshall rarely saw the town he helped layout. His home was in New York and that is where he preferred to be amid the hustle and bustle of the big metropolis. It was where he could bid for large coal contracts amid the businessmen whom he met with daily. He would win a contract and send it on to Shallmar to be filled. He never had to break a sweat or get his hands, or any other part of his body, dirty.

He left the day-to-day work of filling the contracts and running the Wolf Den Coal Company to his mine superintendent who, for all intents and purposes, was also the mayor of Shallmar. The town had no municipal government or mayor, but the coal company ran the water system, generated power and owned the land.

Laying out the town probably wasn't too hard seeing as how the usable land along that stretch of the North Branch Potomac was limited by the river to the south and the steep incline of Backbone Mountain to the north. Shallmar, which came from switching the syllables in Wilbur Marshall's last name, had one street that was shaped like a fishhook with a few barbs on it.

That's probably a more-than-appropriate way to describe Shallmar because just like a fishhook is hidden beneath the bait that lures the fish to bite the hook so that it can be caught, Shallmar's trap was hidden beneath its attractive appearance. Once miners, or diggers as they were often called, were lured in, they would be hooked, finding it hard to leave the small town. Once you lived in Shallmar, other coal towns would feel like a step down. Shallmar was a town where miners could live in houses that were as nice as those of mine superintendents in other coal camps.

Now Shallmar wasn't the only coal town that trapped miners. Lots of coal-mining camps did it. Sometimes it was the only way for a coal company to keep men working in dangerous conditions for little pay. Diggers needed a reason to risk their lives every day from fires, rock falls, explosions and other dangers found in coal mines. In 1923, for instance, five miners in Western Maryland were killed in mining accidents while working for less than $100 a month or the equivalent to $15,508 a year in today's dollars. Between 1900 and 1930, an average of 2,210 miners were killed each year in the United States.

Sounds a little crazy, doesn't it? Especially when you consider that Maryland miners tended to be better paid and work in better conditions than other miners. Miners definitely needed extra incentives.

Different towns used different types of bait. For some it might be higher wages, which diggers might find out were offset by higher prices at the company store. Others might be quick to pay wages in company scrip – most miners called them flickers. They were only good at face value when used at the company store; otherwise, it was discounted ten to fifteen percent. Since the company store tended to overprice its merchandise by the same amount, miners wound up overpaying no matter where they spent their flickers. The Wolf Den Coal Company paid its regular wages in cash, but if a digger needed credit, it was paid in flickers.

Take Dodson, for instance. It was the coal patch just to the south of Shallmar. You couldn't drive there from Shallmar, but you could walk across a large culvert at the south end of Shallmar to reach the town. Dodson had roughly the same population as Shallmar, depending on when you counted heads, but Dodson's trap was that the Garrett County Coal and Mining Company provided a variety of things to encourage its miners to spend their flickers and helped keep them poor.

For Saturday's sinners, a town building housed a bowling alley, pool room and doctor's office. Not only was the doctor good at taking care of most minor injuries, he could also handle any injuries from fights that might break out in the pool room to more-serious injuries if something bad happened in the mine. The latter happened quite often compared to other types of work. On the building's second floor was a "theatre" that was used for shows, dances or any other event where a lot of people gathered.

Then for Sunday's saints, Dodson had a Methodist church. It didn't matter whether you were Catholic, Baptist or even Jewish, if you wanted to hear preaching on Sundays, you had to take it Methodist style. Finally, Dodson also had your company town basics of a store and school.

Meanwhile, there was Shallmar looking all bright and pretty across the culvert, but that was about it. Shallmar had houses, the mining buildings, the company store and a school, which it managed to steal from Dodson because Dodson also happened to be dying out as Shallmar was developing. The town even had rows of garages near the company store and further down the street toward the center of town. Each house was paired with one of the garages. Miners could use their garages for storage, and in the rare instance when a miner might own a car, parking.

As one coal miner put it, "Mr. Marshall put a lot of up-front money into the town before any coal was ever mined."

So, as you can guess, Shallmar was a sharp-looking town as far as mining towns went. Even today when the name Shallmar comes up, people remark on how nice looking it was, which is kind of amazing since no one still around ever saw it in its heyday. But you go ahead and ask someone living in Western Maryland coal country, "What about Shallmar?" They will probably reply, "It was a good-looking town."

Flickers or company scrip were company-issued currency valid only in the company store. They could be bills or coins like these coins from the Shallmar Company Store. Photo from the author's collection.

Of course, it didn't matter that mine runoff and other pollution killed nearly anything alive in the North Branch Potomac. Not that most people realized that it was acid mine drainage from all the coal mines along the North Branch Potomac causing the problem, and truthfully, since many people used the river as a place to dump their garbage, they probably wouldn't have cared.

Most coal camps were built with the knowledge that they weren't going to be permanent towns. They were built quickly and cheaply, and when the coal played out, they could be torn down just as easily. Coal companies also tended to spend as little as possible on housing in order to reserve capital for mine improvements and equipment.

Wilbur Marshall took a lot of care in planning out a town that was built to last. He sent the Italian stonecutters who had built his New York home to Shallmar to construct the powerhouse, the company store and the foundations for the company houses in Shallmar.

Shallmar's houses were built larger than many of the house styles in other coal towns. There were two-story single houses, two-story double houses, bungalows and even a boarding house for single diggers. They were all constructed with a chestnut frame and German siding on top of a cut-stone foundation.

Each summer Wolf Den Coal Company maintenance workers gave the houses and trunks of the silver maple trees in town a fresh coat of whitewash. That may sound excessive, but it really wasn't. Coal dust turned the whitest white to a grimy gray in just a few months. Washing the houses took a lot of time and it wasn't completely successful. If you scrubbed hard enough to get the coal off the house, some of the paint was sure to come, and that just wouldn't do in the town that was supposed to be the prettiest in the region. Whitewash was cheap enough to apply each year plus it protected the trees. With a fresh coat of whitewash, the town seemed to shine as it stood out from the green trees, brown mountains and orange river around it. By contrast, other towns blended in with their surroundings as if to show they didn't plan on being around long and could be easily forgotten.

Another attractive feature of Shallmar was that it had running water. Not just any water, either. The water in Shallmar was soft, sweet and cold. More importantly, it was free from any mine contamination. Wilbur Marshall had a pipeline laid to bring water from mountain springs down into the town. It was piped into a few houses, but otherwise, it came out at a number of spigots throughout the town.

Residents would fill buckets from the spigots to fill a tub or sink. Sometimes, they would connect a hose to the spigot and water their gardens. It was a little effort, but the alternative was to draw water from wells using a large hand pump like they had to do in Dodson. And it wasn't too long before diggers in Shallmar began tapping into the water lines to bring the water directly into their houses.

Yes sir, fine city living had come to coal miners of Western Maryland, but if Wilbur Marshall had really wanted to impress miners, he would have included a commode in every house. Coal companies shied away from building houses with indoor toilets. Not only did they see it as an unnecessary capital expense, but maintaining them could also become costly.

A coal company official in West Virginia once said, "I think I had at one time twenty substations, about thirty-five locomotives, 100 mine machines and 200 stationary motors and looked after about nine mines at once. Those bathrooms gave me more trouble than all the rest put together."

Indoor plumbing was rarely seen in Western Maryland coal patches until the 1950s. Many people had only heard of a toilet, though they had never seen one.

Kenny Bray, who was a Shallmar digger for many years, would sometimes tell a story about his brother Bob from when the Brays lived in Nethken, West Virginia, in the 1920s. Nethken was a coal town just across the river and up the mountain from Shallmar. It seems that when Bob was nine or ten years old, he won a prize at school. He would get to spend the weekend in his teacher's home. Miss McGee stayed in Nethken during the week, but on weekends she lived with her family in Piedmont, West Virginia. It was a town on the North Branch Potomac but about three and a quarter miles northeast, as the crow flies, and about four times that distance if you were driving.

Now a weekend at your teacher's house might sound like a punishment more than a prize, but you have to remember that coal-mining families lived in small houses that generally had only two bedrooms. The parents had one of the bedrooms and the other bedroom was for the kids and many families had four or more kids in those days. The Brays had five children while they lived in Nethken. This generally meant that some of the kids found themselves sleeping on cots in the family room. With so little privacy a chance to stay in a guest room at his teacher's house was like winning a weekend at a four-star resort would have been for other folk.

However, Bob's mother knew that her boy had never seen an indoor toilet, much less used one. So she went around the town until she found a Sears and Roebuck catalog that hadn't been ripped apart for toilet paper. She opened the catalog and showed Bob the pictures of the indoor toilets and explained to him how to use one. After all, she

13

didn't want her son to come across as a country hick to his teacher. The houses of Shallmar all had outdoor privies. The federal government's Works Progress Administration eventually improved these original buildings during the Great Depression, which made Shallmar more attractive to families. The original outdoor privies were attached to the coal or wood shed and there was no pit under them to collect the wastes. Instead, it was collected in a bucket or box set underneath the hole and someone would have to take the bucket out when it was full and dump it in the brush or the river.

That made for a lot of bad smells and unwanted company around the houses. Rats were often seen rooting around the privies and flies were ever present when the weather was warm. Residents hung multiple strips of fly paper near their doors only to see them fill up in a week or so. And this was in a town that was considered a desirable place to live. Imagine what the smell was like in other towns.

Then the WPA came in during the 1930s and built everyone a brand new privy with a lined pit underneath it to capture the wastes and a pipe to the roof to vent the smell. Everyone was so proud of these buildings that they started calling them "Roosevelt Monuments" in honor of President Franklin D. Roosevelt who created the WPA, which was one of his New Deal "alphabet agencies."

One of the other perks of living in Shallmar was that you could get cheap electricity when many towns had none at all because coal camps tended to be in isolated areas. The Wolf Den Coal Company didn't go out of its way to provide this for the diggers and their families. The company needed a power plant to run the mine ventilation fan and some of the other equipment in the mine so it built its own power plant. The Shallmar powerhouse had two 21-foot horizontal, coal-fired, steam boilers; a Ridgeway 450-horsepower steam engine with a 250-volt generator; a Skinner 200-horsepower steam engine and generator; and a steam turbine and generator. The water for the boilers was piped through the mountain in the company's water lines, which is the main reason that the town had its own water system.

Selling excess power cheaply helped offset some of the company's costs to run the mine. If you lived in Shallmar, you had electricity and you paid for it whether you used it or not. A monthly charge was deducted directly from the miners' pay.

While cheap power sounded attractive to women who had to hand scrub clothes that were stiff and black with coal dust or had rooms that smelled of burning kerosene, the electricity was direct current.

Most of America used alternating current to run their home appliances. Plug a washing machine into a direct current outlet or a light bulb into a lamp connected to direct current and neither one lasted too long if they even worked at all.

One of the generators at the Shallmar power house. Photo courtesy of the Western Historical Library.

You needed appliances specifically made for direct current and they weren't too easy to get, particularly when most of Shallmar's diggers could only shop at the company store. The Shallmar company store carried direct current light bulbs, but most people considered them too expensive and stuck with kerosene lamps. The company store also carried 250-volt irons and washing machines. The washing machines were gasoline or regular AC electric washers that had been converted to DC power with a 250-volt, ¼-horsepower engine.

To find bigger items that ran on direct current, you had to go a city and shop for it, but since few miners had a car, this rarely happened.

The other problem with using power in Shallmar was that you could only get it when the mine was in operation. Since the mine rarely operated seven days a weeks, that meant you were without power on some days. What good did it do have a refrigerator or washing machine when it couldn't be used for a good portion of the day? Most people opted for an ice box and wash tub because they knew they could be depended on to work.

And even on days when the mine was in operation, the electricity

15

only stayed on until midnight. At a quarter to twelve, the maintenance supervisor would cause the power to flicker twice to warn miners and their families that power was about to shut off for the night. Once it was turned off at midnight, it didn't come back on until the maintenance crew started the generators before the miners came in for their workday.

Each house along the main road in Shallmar had hedges planted in front of it and a trellis over the front door entry that was covered by red rambling roses. All of this was maintained by the mining company.

Then just to show off a bit, on the river bank across the street from the company store, the maintenance men kept the lawn nicely mowed. Amid that field of green, they placed whitewashed rocks so that the rocks spelled out "SHALLMAR." This was for the benefit of passengers on the daily trains that stopped at the Harrison, West Virginia, station across the river from Shallmar.

You could cross from Harrison to Shallmar on a swinging bridge, which was made of wooden planks on cables suspended across the river. Young boys liked to jump on it to show how the bridge had earned it name, but most people just used it to go from town to town and were very happy if the journey was a relatively still one.

You wouldn't find anyone who lived in Shallmar complaining too much about the inconvenience of intermittent electricity, direct current or having to go outdoors for water. They didn't even consider them inconveniences. Miners' wives were excited to have a trellis entrance and the diggers were happy to have rooms that didn't make them feel like they were still crawling through a mine shaft. However, you can see how mining companies fed off their own miners, finding ways to take back their earnings without offering an equal value in return whether it was overpriced goods, electricity that couldn't be used well or water that would have been produced anyway.

Shallmar looked so pretty and had so many features that were attractive to people who typically did without a lot that you needed a reference to rent one of the bigger houses on the main street.

Remember, how I said there were barbs in the fishhook shape of Shallmar's only road? These were little cul-de-sacs that ran off the main street to a cluster of small bungalow homes. They were still nicely kept, but they just weren't as fancy as the ones on the main street. They were also cheaper to rent. Since miners were notoriously underpaid, unless they had large families, most of them wound up renting houses off the main street because it was all they could afford.

Families of Italians and Poles self-segregated and lived in the bungalows in an area behind the company store. The families had come to Shallmar to help build the spur rail line into town and then stayed to work in the mines.

Kitzmiller was always a dry town, but Prohibition made the sale of beer, wine and liquor illegal everywhere else, too. That didn't slow down Western Maryland diggers. When they could no longer buy their preferred liquid refreshment, they made moonshine, home brew and pick handle to replace whiskey, beer and wine. Pick handle was a homemade wine made from fruit, grain, sugar, yeast and water.

Moonshining was a profitable business. For about $1.05 in ingredients, 5 gallons of home brew beer could be made and sold for 25 cents a quart or about $5 a gallon. You had to be careful who you bought from, though. Some moonshiners took shortcuts to make their concoctions faster or to get more from a batch. For instance, one shortcut involved putting rubbing alcohol into the corn mash used for to make moonshine to increase the amount that was produced. Your first taste of that brew could put you in the hospital.

Shallmar saw good years during the 1920s. It was still a new coal patch and well maintained; the jewel of Western Maryland. The Wolf Den Mine produced more than 100,000 tons of coal every year but three during the Wolf Den Coal Company tenure. I guess it could be said that Shallmar's golden years were those when Wolf Den Coal Company was in business...with two exceptions.

During the last week of March 1924, warm temperatures began melting the snow on the mountains. March 28 dawned with an overcast sky. The temperatures were warm and a light rain fell throughout the day and into the night. Both the temperature and rain helped even more snow melt.

As residents along the North Branch Potomac watched the water rise, they realized there might be a problem. Around 9 p.m., a night watch was posted on the river in case it overflowed its banks. At 2 a.m. in the morning of March 29, the night watch sounded the alarm. Similar alarms were sounded in communities all along the river.

People in Shallmar, Kitzmiller and other communities threw on clothes, grabbed what they could and ran for higher ground. Lanterns flashed, swinging back and forth as people hurried out of their homes into the pitch dark of the night. Parents called to their children to try and keep them from getting separated in the darkness.

Soon enough, people heard the sound of rushing water where there shouldn't be any. As dawn broke, they could see chicken coops with chickens in them, debris, livestock and even houses rush by in the flooded streets of their towns. One house in the North Branch floated by with a light still burning in an upper window. It was only extinguished when the house smashed into the Kitzmiller bridge and became debris. The flood eventually swept away the bridge, too.

Samuel Beeman, his wife, Beeman's father and the Beemans' two children climbed onto a tall tree when the flood waters got too high in Kitzmiller and began filling their home. The raging river eventually washed away their house. The water pushed against the tree they were on while at the same time carrying away soil from around the roots. The tree uprooted and toppled into the river taking all the members of the Beeman family with it.

Another man was last seen wading on Main Street in Kitzmiller before he turned up missing and was presumed drowned. Other people lost livestock. The flood waters also took out the tipple at the nearby Pee Wee Mine and thirty loads of coal, roughly sixty tons, were washed into the river near the Potomac Manor Mine when the coal cars were tipped over by the floods.

The flood began receding around 11 a.m. and people came down from the high ground to see what was left of their homes. Towns all along the North Branch Potomac felt its affect. Cumberland had $4 million in flood damage. Closer to home, Chaffee, the next small coal-mining town downriver from Kitzmiller was so badly damaged from the flooding that it was abandoned and the miners moved to Vindex. In Kitzmiller and Shallmar, twenty-one houses plus a number of cars and trucks were either destroyed or washed away.

Even after the North Branch Potomac had returned to within its banks, some of the families whose homes were too damaged or washed away spent several nights sleeping in the Shallmar powerhouse. It was large enough to house them all and high enough up the mountain that it hadn't flooded at all.

The cost of recovering from the 1924 flood contributed to the failure of the Wolf Den Coal Company in 1927 and its restructuring as the Shallmar Mining Company with the same officers.

The new company did well as it moved towards its peak production year in 1929, but the downturn in overall coal demand hadn't changed. Coal had been losing ground in the energy market since 1910 as oil, natural gas and electricity became more accepted.

The second exception to Shallmar's golden years happened on November 11, 1930. Early in the morning, the fireman at the power-house saw lights where there shouldn't be any. He tracked it down and saw the tipple was burning. He sounded the mine whistle to signal for help putting out the fire. The mining company had its own fire department equipped with a reel and hose stored at the powerhouse that could be pulled to where it was needed and connected to the water system. However, by the time the miners arrived to help, the fire was out of control. The tipple burned to the ground for an estimated loss of $28,000.

The problem with coal is that once it is hauled out of the mine, it's gone. Given enough time and resources, it will all disappear.

Getting that coal was the trick. How the diggers mined the coal worked depended on the coal seam and how thick it was. It also depended on how much the mining company was willing to spend, and let me tell you, they wanted to spend as little as possible. That's one reason mining could be so dangerous. Not only was the mountain itself unpredictable, but a company that cut corners on safety was betting the lives of miners against the stability of the mountain.

Shallmar's Wolf Den Mine was considered a safe one. The mine had three openings that led into the mountain on a slight incline, which was by design so that water that would filter through the ground into the mine would run out rather than pool near the face.

One opening was for air ventilation and the other two were used for getting into and out of the mine. Two 15 mile per hour fans supplied the mine with 100,000 cubic feet of fresh air each minute. It had a double drift entry system, which means that miners could walk right into the mine rather than taking a lift car down into the mine shaft. In other words, drift mine entrances were dug horizontally into the mountain while deep mines had entry shafts that went vertically into the mountain. Drift mines also have their shafts dug so as to follow the coal seam rather than trying to cut across it like crosscut mines.

Inside the Wolf Den Mine, it used what's called a "room and pillar" system. It was the preferred techniques for coal mines back then. With demand for coal falling off, it didn't matter much if it was also a wasteful technique because it was inexpensive.

Unfortunately for miners, it was also a dangerous way to dig coal. Coal was mined in "rooms" that were about 100-feet wide so that a wall of coal and rock was left intact between rooms. These walls or "pillars" served as supports for the tunnel roof.

Once the edge of the coal company property was reached, the mining operation would essentially back itself out of the tunnels, mining the walls as they went. This would eventually cause cave-ins in the rooms as the weight from the mountain pushed down.

Miners ride a coal car out of a mine. Courtesy of the Western Maryland's Historical Library.

Two miners would work this part of the operation because of the danger involved. With each succeeding wall that was removed, the pressure grew greater on other walls. Removing the final walls in a tunnel was particularly dangerous because the pressure could cause a weak wall to explode turning rock and coal into shrapnel or the tunnel could collapse prematurely crushing anyone in the room. Removing walls took the lives of both experienced and inexperienced miners.

Now Maryland mines were generally considered safer mines to work in. Whether that's by design or chance is hard to say, but fewer miners tended to be killed in Maryland mines.

This doesn't mean that there weren't accidents in the Wolf Den Mine. Each year, miners suffered from contusions and broken bones, but the Wolf Den Mine didn't have its the first fatal accident until 1932, and technically, that one didn't happen until 1933.

John Melouse, his son, Charles Rozier and his son were retimbering a section of the mine on December 28, 1932. The Melouse family had emigrated from Yugoslavia in 1905 to earn their piece of the

American Dream in the coal mines. At that time, their last name had been Milavic, but when they entered the United States at Ellis Island, the immigration clerks couldn't understand their heavy European accent and wrote Milavic down as Melouse.

While the group of miners was eating lunch that day, John Melouse noticed that some of the timbers needed trimming so he went to the face to get his saw. A rock fell on him while he was on his way back to the group. He died from his injuries on New Year's Day 1933, leaving behind a wife and five children. Though the accident was preventable, the Maryland Bureau of Mines didn't fault the coal company because precautions had been taken and the miners were experienced. It was called an "unusual" accident. "It could hardly be anticipated that the rock on the brow would have broken the way it did," the Bureau of Mines report read.

John Melouse's family probably would have called the accident "heartbreaking" or "life changing," but no one ever accused the Bureau of Mines of being sentimental.

When an accident happened, just about everybody in town saw the hearse drive up to the mine and they knew what it meant. They would drop what they were doing and rush to the mine. The women would be nearly hysterical from the stress of waiting to see who had been killed.

"I remember my mother; she'd almost go crazy when she saw a hearse go up through town. It was really sad," George Brady said once.

Usually the hearse would take the bodies across the river to Blaine, West Virginia, because that was where the nearest undertaker worked. The mine would shut down the day of the funeral. The company probably would have kept it open if they could, but it would have been hard to get enough people to work. They would have all gone to the funeral.

After that, it was only a matter of a couple weeks before the miner's family moved out. Without a miner working for the coal company, there was no reason for the family to remain in town and no one was paying rent or for electricity and coal, all of which were deductions made from a miner's pay. Also, the company officials were more than happy to see the family leave town so as not to be constantly reminded of how quickly the mine could take a life. Such was life in a company town.

The work pressed on. It had to.

In the late nineteenth and early twentieth centuries, coal powered the country. It turned the turbines on the great ships at sea that made up the U.S. Navy. It burned to such a degree that it melted iron in the furnaces so that steel could be forged. It created steam to move trains. It warmed the houses in the coldest reaches of the country. Without it, the country's national defense faltered, large-scale construction stopped and families froze.

Coal camps like Shallmar, while not glamorous like New York or Chicago, were just as important to the country.

John L. Lewis saw to that.

He became the president of the United Mine Workers of America in 1920 a couple years after the Wolf Den Coal Company began operation. His election didn't matter much to mining company officials in Maryland since the UMW had little sway in the state at the time, but Lewis was more of a force of nature than a man. When he set his mind to something, nothing stopped him. Luckily, he was on the side of coal miners.

He had thick, dark eyebrows that drew your gaze to his eyes and once his eyes locked on yours, they held you like a ferret that bites your finger. Then his voice would boom and you might be forgiven if you thought it was the voice of God. The only difference was that God smiled; John L. Lewis rarely did.

Lewis realized coal's importance to the country and used it as leverage to strike for higher wages even in the middle of World War II. That was an unpopular strike that even the miners didn't like. For the same reason that Lewis believed he could use a strike to get higher wages for union miners was the same reason the strike failed. Yes, coal was important to the country and the war effort. In fact, it was so important that mining coal was seen as a patriotic act so striking was seen as unpatriotic.

When negotiations failed, President Franklin D. Roosevelt told the miners and the country in a national radio broadcast, that "every idle miner directly and individually is obstructing our war effort." He had the federal government take over the country's coal mines and he sent soldiers to them to ensure that the flow of coal continued unimpeded.

Unlike the 1922 coal strike that saw plenty of violence and killing, no violence was generally reported with this strike. The only change that area mines saw were that they started flying a U.S. flag and some of the mines posted the legal orders that gave the government authority over the mine on bulletin boards at the mines. Some of the miners would even salute the flag as they came to work.

That doesn't mean that officials weren't on the lookout for union violence. Kenny Bray was working in the Wolf Den Coal Corporation at this time. He couldn't afford to buy himself a watch so that he would know when his work day was over so he carried an old alarm clock with him. He kept it in his powder bag along with his explosives he needed to shoot coal for the day.

One morning, he filled his powder bag for the day and then set his clock inside on top just like he had done for weeks. Then he put his bag in the pile along with the other diggers' bags. The miners walked to the place where they usually waited for a coal car to take them into the mine. This particular morning, the men were told that a coal car had derailed inside the mine and it would be awhile before it would arrive so they should go ahead and walk to their work places, which were about a half mile under the mountain. They cold drill their holes in the face and by the time they were ready to shoot the coal, the powder car would be able to deliver their bags.

As Kenny would tell the story, "A while later Roy came to my work place with my blasting powder. He told me that they had started to put the car on the rails and heard the clock ticking. They didn't know about the clock. They thought someone put a time bomb in with the explosives and they were afraid to go near it."

It wasn't until another motorman came along and remembered that Kenny carried an alarm clock with him that the crew at mine entrance would approach the powder car.

"So the scare was over, but I was told not to put the clock in with the explosives in the future," Kenny said.

The fact that there was no violence seen during this strike is probably because the miners wanted to work and couldn't with the union was keeping them out of the mine. The armed soldiers around the mines probably discouraged any violence, too.

The miners would call out to them sometimes, "You can't dig coal with a bayonet!"

The government even had an American flag flying over the mines that they took over. If the miners or union had retaliated, they would have been seen as fighting against America. So there was no gunfire, luckily, but the soldiers were armed as soldiers are wont to be.

3

A Farmer's Son

Yes, Junior Paul Andrick's real first name was Junior. It wasn't a country hick thing, though he didn't know why his parents had named him something at that beginning of his name that belonged at the end of his name. He rarely even went by his first name except in official circumstances when he needed to use his legal name.

Paul grew up in Philippi, West Virginia, where he had learned how to mine, after a fashion. Philippi was coal country. The coal mines and Baltimore and Ohio Railroad were the major employers in the area.

He wasn't a coal miner, though he squinted a bit like coal miners did in daylight. The difference was that Paul stopped squinting when he put on his glasses. Miners stopped squinting when they got out of the light of day.

The Andricks were farmers. Arley Andrick, Paul's dad, never had a need to hire summer help for the harvest. That's because once Arley and Dora got going on their family, they didn't stop until they had nine boys and a girl. Paul was number four born in 1916 about a year before Shallmar was built.

Now the farming life can be a hard one like coal mining is, but it's a different kind of hard. It was a good kind of hard. Farmers worked long hours, sometimes in harsh conditions, but it's not something they could run to a union and complain about because they worked for themselves and not a company.

If a miner worked, he got paid each day for the amount of coal he sent to the tipple. A miner's challenge was to keep working as the market for coal faded. During Paul's adult life coal miners were lucky if they could work three days a week. Farmers, though, could work from dawn to dusk and then some, but his payday wasn't until the harvest came in. Then bad weather or poor crop prices could mean a mighty lean winter.

That happened to the Andricks more than once, but they never went hungry and Arley and Dora had twelve mouths to feed and three-quarters of them were growing boys who seemed to need to consume their body weight in food each day.

With such a large family, Arley worked a second job as a carpenter in the coal mines around Philippi for twenty years. The extra income helped the Andricks make ends meet during lean years and it gave Arley work in the winter. It worked out well because farmers had little to do in the when coal demand was greater and there was more work at the mines. Arley's mine work also led him to become a union advocate and organizer in the 1920s.

The difference between farming and mining, as I see it, is that farmers, by the nature of their work, have to be planners. They have to plan what crop to plant where and how much of each and estimate a market need that is months away. Then they have to estimate what their yield will be and how much it might fetch at market in case they need to go to the bank for credit and when they try to sell their crops before it's harvested. Then whatever they earn from selling their harvest, they need to make stretch out to last until the next harvest.

Then there's also the matter of holding some of your crops back for your own use. When crop prices are low, you hold more back and do a lot more canning. When prices are higher, you sell more and eat some of what you canned.

Oh, there's a heap of planning in farming. People think farmers go out each day in the fields and do hoeing, planting and weeding with nary a thought running through their heads. That's not the case at all. A lot of farmers, the good ones at least, spend that time studying the land, thinking about the future and planning.

If a miner did all that in a coal mine, his last thought would be about what a waste of time it had been because a miner thinking about anything other than mining didn't live to enjoy the future he planned.

Paul grew up with a strong work ethic that his parents instilled in him and his brothers and sister. With such a large family, everyone was expected to do his or her part in order to keep everyone fed and in their house. While his father believed that farming was honorable work, he knew it could be chancy and hard.

Like most parents, Arley and Dora dreamed of a better life for their children and they believed that education was the key not only to planning crops but planning lives. So along with having chores around the farms, the Andrick children were expected to do their best in school.

Of the ten Andrick siblings, six managed to go to college; not an easy feat for a farming family to pull off.

In Paul's case, he attended Fairmont State Teachers College in Fairmont, West Virginia, on a football scholarship. Although he didn't look it at first glance, Paul was pretty solidly built. He stood five feet nine inches tall and weighed 160 pounds. A newspaper article from 1938 described him as one of "State College's little, but mighty, guards."

In those days of football, the emphasis was not so much on mass. You really didn't want to be tackled too much back then. Football had been in danger of being banned on college campuses in the early years of the twentieth century. It had little in the way of protective gear and the rules allowed for dangerous plays. Between 1901 and 1904, forty five young men were killed playing high school or college football. It took a whole lot of people who didn't want the sport to die, like President Theodore Roosevelt, getting together to make if a safer game to play. The group of reformers agreed on rule changes and protective equipment had allowed the sport to grow in popularity and safety.

Paul didn't escape injury playing, but he couldn't blame his problems all on football. A life of hard labor and bone-jarring recreation had contributing to him developing arthritis. It wasn't bad enough to cripple him, but it did begin to slow him down somewhat as he got older.

Since Paul only had a partial football scholarship, he also needed part-time jobs to make ends meet and make up the difference in his $30 a semester tuition that his scholarship didn't cover. Work was hard to find and didn't pay much. He took what work he could find. When he wasn't in class or playing football, he was hauling logs for a local company or stoking the furnaces of the girls' dormitory with coal. As part of his college work in the girls' dormitory, he would also mow the lawn or shovel snow when either was needed. For his efforts, the college provided him with a cot and desk in the furnace room where he lived on campus.

Both jobs involved heavy labor, which didn't do his arthritis any good, but the work helped keep him in shape. Plus, working as a fireman certainly helped his social life. Although he didn't have a lot of time to date, he did have plenty of opportunities to meet girls who lived in the dormitory. He had to stay in the basement while he worked, but he still had chances to meet the women of Fairmont State while going in or out of the basement door. He would stop and talk to them and the fact that he lived and worked in their basement was a good conversation starter.

While he worked, the door from the basement into the dormitory was kept locked so there was no inappropriate fraternizing, but he could hear the women talking on the first floor. At times, Paul could feel his body heat up and he couldn't be sure whether it was hormones or just the heat from the furnace.

Martha Jane Andrick

However, it wasn't at the dormitory where Paul met Martha Jane McDonald. He saw her at a school dance in the gymnasium and asked her to dance. While they swayed to the music, they talked. Then they danced some more and talked some more. With each song, it became harder to keep the required amount of light between them.

Martha, or Molly as Paul called her, was an education major like Paul and she also hailed from a small town. In Molly's case, it was Mayville, West Virginia. She said she had a large family, but from Paul's perspective, Molly's three sisters and four brothers added up to just a medium-size family. After all, he had nine brothers and a sister.

Molly and Paul began dating, but with two years between them, it seemed that the romance might be short lived. Molly doubted that she would ever see Paul again when he graduated in 1939. After all, he had returned home to Philippi and started looking for work as a teacher.

Then one day that summer, Paul showed up on the porch of her family's house in Mayville, which was eighty miles from Phillipi.

"How did you get here?" Molly asked him since she knew that he didn't have a car.

"I hitched a ride. It took a while, but I had to see you again," Paul told her.

And so love won out.

The two of them were married in 1940 and it was only then that Molly learned that Paul wasn't actually named "Arley" like his father. She watched him fill out the marriage license at the courthouse and write "Junior Paul Andrick" on the line for his name.

"Your first name's Junior?" she asked, surprised.

"You knew that."

Molly shook her head. "I thought your name was Arley."

"Why would you think that? I told you my name was Junior."

"I thought that was a nickname because you were named after your father."

They shared a good laugh over it, but it wasn't the last time something that should have been obvious to them wasn't.

Paul and Molly were married in small ceremony, not because they wanted to keep things inexpensive, they just didn't want it publicized that they were husband and wife.

Don't get the wrong idea here. It wasn't that Paul and Molly were ashamed of each other. You only had to watch them for a couple of minutes to know that to be false. No, the problem was that in those days, at least in West Virginia, if Molly was going to teach, then she needed to be single. (All you women reading this can express the appropriate amount of outrage now.)

It looked like West Virginia was going to be where the newly-weds worked, too. Paul had graduated with his bachelor's degree and gotten his state teaching certificate. Molly had gone a different route in order to graduate at the same time as Paul. She had earned enough credits after two years at Fairmont State Teachers College to qualify for a Normal School Certificate. It allowed her to teach in West Virginia public schools, which in some cases meant that she might be only two years older than the students she was teaching. West Virginia was the only place she could teach, though. Any surrounding state required a bachelor's degree in order to get a state teaching certificate.

Martha and J. Paul Andick in the home in Shallmar. Photo courtesy of Jerry Andrick.

Married life suited the newlyweds. Paul was hired as an elementary school teacher after he and Molly were married and they settled down to domestic bliss. Molly was worried, though. Paul hadn't rushed off to join the military after the Japanese attacked Pearl Harbor, but there was always a chance that he could be drafted. He had registered along with millions of other men in 1940 before he was married.

Paul didn't let it bother him. He continued his work, transferred to another elementary school and then to Bayard High School to teach math classes.

Along the way, the Andricks also had their first child in 1942. Just to really make the family names confusing, the baby was named Jerry Paul Andrick, another J. Paul Andrick like his father, but this J.

Paul wasn't a "Junior" at either end of his name.

Paul was still teaching at Bayard High when his number came up both figuratively and literally. His Selective Service Number was drawn and he was drafted into the U.S. Army in March of 1945 for the duration of the war plus six months. He shipped off to basic training at Camp Blanding in Florida.

The army camp was located about 45 miles southeast of Jacksonville. It had been created in 1939 for the Florida National Guard, but when the U.S. entered World War II, it moved from state to federal control. The federal government leased 170,000 surrounding acres in order to enlarge the camp's original 30,000 acres so that it could hold two infantry divisions and other units of the U.S. Army.

During the course of the war, the 200,000-acre camp also housed a 2,800-bed military hospital and German prisoner of war camp. More than 800,000 soldiers would be trained there.

In 1943, it became an Infantry Replacement Center where draftees were trained and sent overseas to replaced infantrymen who had been killed or wounded in the fighting. By the time Paul arrived in 1945, a large portion of the infantrymen being sent into the war trained at Camp Blanding. When the war ended, the camp also added a separation center.

Since Paul had to leave with school still in session, Molly stepped in and taught her husband's classes until the end of the school year. As soon as classes finished for the summer, Molly headed down to Florida to be with her husband until he shipped out to fight in the war. Three-year-old Jerry stayed behind with his grandparents in Philippi.

Paul completed his Basic Training, but not without some difficulty, which surprised his training officers since he appeared to be in shape. The doctors re-examined him and determined that his football injuries and arthritis limited what he would be able to do as a soldier. Paul was discharged and was listed as a disabled American veteran.

4

Shallmar Declining

Change came slowly to a miner's life. Some might say it was because the diggers were too much like the mountains they mined. The mountain could only be moved when the coal and rock was removed from it bit by bit. The same was true for miners. They could only be moved bit by bit.

Coal was initially dug from the ground with a pick and shovel and it continued that way long after coal-cutting machines were introduced in the 1880s. Only over decades did their use become more and more accepted. The isolation of mining communities meant that innovations appeared slowly – indoor plumbing, electricity and telephones, to name a few.

On the one hand, this isn't too surprising because mining towns were in rural areas, which also saw improvements come slower than they did in cities. On the other hand, many mining towns could only dream about these innovations even after other rural communities already had them. For instance, telephones were in common use throughout the United States in 1949, but in Shallmar, there were only two – one in the company store and one in Mr. Marshall's house. The same was true for bathrooms. Shallmar had around ninety houses, but only four of them had indoor bathrooms in 1949.

On a work day at the Wolf Den Mine, a line of diggers started forming on the main street shortly before 6 a.m. each morning. The miners trudged south rounding the fish hook turn and up the mountain to the entrance to the mine. When it was winter and still dark that early in the morning, the miners looked like a line of giant fireflies. The men would turn on the lamps on their helmets, whether they were carbide or battery powered, so they could see the path to the mine in the dark. With no street lights in town, Shallmar was pitch dark at that hour in the winter and their head lamps bobbed along in a jumbled line.

These men wore no suits for their work. Their work clothing was overalls or jeans, leather gloves, hard hats and heavy boots or steel-toe safety shoes that protected their toes if rocks fell. Many of them wore sweaters or flannel shirts even in the dog days of summer because it was always cool in the mine. Whatever they chose to wear, it was heavy duty so it would hold up to being scraped on rocks and rubbed along the ground. The men also carried their number-four shovels and large dinner buckets with them to the mine.

Miners enjoying a lunch break together. Courtesy of Western Maryland's Historical Library.

They would stop at the powder room and fill a bag with the black powder they would need for the day. They put their shovels, lunch pails and powder bags into a coal car that would carry the equipment to the layoff where the miners would pick their items up before they headed to the face.

The miners would then walk back to the shop where the equipment was repaired, tools sharpened and ponies shoed. They would hang their miners' checks—brass medallions about the size of a quarter and stamped with their miner number—on a board outside the shop. Then they would make sure their helmets were seated securely on their heads, turn on the head lamp and head into to mine. If the check was on the board, it meant that miner was underground. It was a safety precaution to make sure no one was left behind in the mine if there was an accident like a cave in. A quick glance at the board would tell the mine foreman whether everyone was out of the mine, and if not, who was still inside. If there were problems in the mine, you didn't want to see a miner's check on the board.

The Wolf Den Mine entrances had a lot of cribbage around them, which was used to stabilize the mountain around the entrances so that rockslides or mudslides wouldn't seal the entrances. The mine's main heading was tall enough that the diggers could walk into the mine standing tall. That changed near the face, though.

If it had been raining a lot, then it would be raining in the mine, so to speak, as rainwater seeped through the mountain and dripped into the mine. It also meant that diggers would be covered with mud as well as coal dust when their shift was finished. It not only got on a miner's clothes, but it worked its way under them mixing with sweat and forming a grit so that it felt like the miner was wearing sandpaper long johns.

However, it was in the spring that miners tended to worry. As the ground thawed and seeds sprouted, the ground tended to move a bit. If you were on top of the ground, well, things were fine, but when you were underground, all it took was a little shift to cause problems like mudslides and roof collapses. The diggers were also responsible for pumping any standing water out of the rooms where they worked, and it had to be done for no pay.

Two coal cars were used to carry men back toward the mine face where the men dug coal from the mountain. The first car carried the diggers and the second car, the powder car filled with explosives needed for the day, usually followed about five minutes behind.

Coal cars weren't designed for passenger comfort. They had no springs or cushions to soften the ride under the mountain. The ride was even worse if the coal car had a "flat wheel." Locking brakes on the coal car could sometimes wear down a section of a wheel so that it wasn't fully round. Each time that section of the wheel hit the rail would mean a jarring bounce for any riders.

The man trip at the Wolf Den Mine went about a mile underground to where the heading split off into lower-clearance tunnels with ceilings that were four-feet high or even lower. The diggers would have to walk bent over or crawl from that point to their work areas at the face where the coal was being mined. If they tried to walk fully erect, they only wound up bumping their heads on the ceiling. You could tell a newbie was in the mine. He would be the man cursing every third step when his hard hat banged on the ceiling.

Miners near the entrance of a mine after a man-trip out of the mine. Photo courtesy of Western Maryland's Historical Libary.

The diggers usually worked in pairs for safety, but occasionally miners would work alone a few hundred yards from anyone else. Once at the face, the diggers might have to spend the day on their sides or knees because of a low ceiling. It was hard work that left the miners sweating. They didn't take off their shirts, though, because the coal dust and dirt would have coated their bodies, plus, the temperature in the mine tunnels was in the fifties.

The only light the miners had was from their head lamps. Carbide lamps, which burned with an orange glow, worked because calcium carbide reacted when wet. The head lamp had an upper and lower reservoir. The carbide was placed in the lower reservoir and water in the upper. By opening a valve, water dripped onto the carbide. The resulting reaction created a gas that would burn providing a bright, broad light. Carbide lamps also had reflectors behind the flames to focus the light forward. Miners would often use their picks to carve a shelf near where they were digging. They could then take the carbide lamps off their helmets and place them on the temporary shelf.

Carbide lamps had been in use in coal mines since the turn of the century. They were slowly giving way to battery powered head lamps, though.

The first step in a digger's preparations to mine coal was to set timbers near the face to give the ceiling some support. Mining was jarring work. Falling rock killed more miners than any other hazard in the mines.

George Brady's grandfather, Elisha Spiker, was one of these people. He was working alone one day in April of 1937 because his work partner hadn't come in for work that day. Someone else should have been assigned to work with him, but instead, the foreman decided to allow Elisha to work alone that day. He was walking through the tunnel when rock broke loose from the ceiling and fell on him, killing him instantly.

Once the timbers in the room were set, the miners would then dig in and lay track to bring the coal cars from the layoff to their room in the mine. The layoff was the point where the track ended and began branching off to the different rooms.

Even though track was laid from the layoff to the face so that coal cars could reach it, electricity wasn't. The electric locomotive engines were used in the main heading while ponies were used to pull the coal cars from the rooms. Diggers were also unable to ride the coal cars from the layoff because the height of the ceilings in the rooms was usually lower than the ceiling in the main heading so diggers were unable to sit up in a coal car once it left the layoff.

Once the track was laid down, the diggers could actually begin digging or rather drilling. They used augers to drill five or six feet into the face. The augers were then removed and a copper needle was inserted along with a black powder cartridge. To save money some independent miners would make their own cartridges by pouring black powder from flasks into a rolled-up newspaper. Black powder cartridges eventually gave way to dynamite sticks and black powder was forbidden. The needle was pushed as far into the auger hole as it would go. Then dirt was tamped in around the needle to fill in any open space. The needle was pulled out leaving the powder cartridge at the back of the hole and a small path open to it.

Knowing where to drill the hole and how much black powder to use was a skill developed with experience. If not enough black powder was used, the resulting chunks of coal would be too large to load in the coal car and the digger would have to break them down into smaller pieces, which would take time away from loading the cars and earning money. Use too much powder and you wound up with too much slack, which was fine coal waste that a miner wasn't paid for. A large explosion could also weaken the ceiling increasing the chances of a cave in or rock fall.

A squib that acted as a fuse was then inserted into the hole. This was a tube of fine powder with a twist of sulfur-impregnated paper on one end that roughly resembled a bottle rocket used by youngsters during the Fourth of July holiday. This fuse was lit and the diggers scrambled to put some distance between them and face.

"Fire in the hole!" a miner would shout as a warning.

The paper burned slowly, but the powder in the squib sent the flame back to the powder charge with a "rocket" effect.

BOOM!

The resulting explosion loosened the coal round the auger hole and allowed the diggers to use a pick and shovel to load coal into the coal car. Diggers would try to separate coal from rock and ash before tossing it in the coal car. It took time away from digging, but it helped the diggers avoid getting docked when the coal was checked outside.

One miner estimated that if a miner swung his shovel five times a minute and weight of each shovelful of coal averaged twenty pounds, it meant that each miner shoveled about 100 pounds a minute. Multiply that by fifty minutes an hour. This allows ten minutes an hour for other tasks like setting a charge and laying track. Multiply that amount by six hours, which leaves time for lunch and setting timbers. The result is that each miner can shovel 30,000 pounds or 15 tons a day.

The problem was that the explosions could make more than just the coal face unstable and they and worsened the air quality in the room. Smoke from the black powder tended to give the miners headaches. Their throats would burn from a temporary lack of oxygen because of the smoke and dust until the ventilation fans cleared the air. To avoid this danger, sometimes the diggers would set their charges at the end of their shift, so the air would be clear for the next day's work.

The Wolf Den Mine diggers sometimes used arc cutters, which were machines with multiple augers that would break up the coal face so that the coal could be loaded. Mining could be done faster with arc cutters, but they were less safe for the miners. The arc cutters ground up the coal into smaller bits creating a fine coal dust that when inhaled. This could lead to black lung disease unless the miners wore face protection, which was not something that was commonly done until the second half of the twentieth century.

Getting black lung disease or "miner's asthma" could be a nasty way to die. The accumulation of all that inhaled coal dust coated your lungs and turned them black so that at the end of your life, you could barely draw a breath. If there was a good thing about hand pick mining, it was that it didn't create as much coal dust as machine mining.

Each coal car could hold two tons. When a car was full, the miner would hang his miner's check on the coal car and have a pony driver hitch one of the mining ponies to the car and pull it out to the layoff.

Ponies tended to be the preferred animal in the mines, though sometimes mules might be used. However, ponies were stronger and smaller. They could go into areas where a mule couldn't. Ponies were also more manageable. All it took to prove that point was for one of the long ears of a mule to hit the charged line that powered the electric locomotives in the main heading. Kicking mules and confined spaces were not a good combination.

At the layoff, the filled coal car was coupled to an electric locomotive engine that ran on the small-gauge railroad track in the mine and hauled the car out of the mine. Once the coal car was emptied at the tipple, the mules would pull the coal cars back to the layoff.

Pony driving didn't require any special skills except to be able to control the ponies by voice commands alone. Reins weren't used because they got in the way. The ponies walked end to end rather than side by side and the pony driver sat on the bumper of the coal car telling them what to do or even singing to them. The drivers would make up their own little ditties to sing to their ponies.

One pony driver sang:

"And now I work in the mines.
I'm a driver without any lines.
On the bumper I sit and tobacco I spit
All over my pony's behind."

Other songs were as dirty as the coal the pony drivers hauled.

All of this work had to be done while the men in the mine watched out for rock falls and black damp or spots in the mine where methane had collected. Black damp could knock a pony down and as quickly as a man and if a pony fell over, it was doubtful it could be moved before it died. It was enough of a challenge moving a full-grown man who had fallen without getting caught in it yourself.

Charles McIntyre was walking through the Wolf Den Mine once with his father on their way to the face where they would be working for the day.

He said, "My dad and I went in, he just went in around the corner from the braddish cloth and I heard him fall. I went around and, of course, had my light and here he was laying there."

Charles grabbed his father's feet and pulled him back the way they had walked; where he knew there was air. Then he began pumping his father's chest and breathing into the older man's mouth. With black damp, methane pushes breathable air out of an area making it a "dead" spot both literally and figuratively for miners. The effect is the same as if you were dunked under water but didn't realize it. With no air to breathe, you pass out, and if someone doesn't pull you out, you die from suffocation with breathable air only inches away.

Just one more reason why you don't want to be alone in a coal mine.

The other problem with the black damp is that it is methane gas. A careless flame or a spark in the wrong place and you wouldn't be worried about breathing; you'd be more concerned about burning. If the gas explodes, you might find yourself under a rock fall.

Outside the mine, the coal was taken to the sorting area near the tipple. The miner's check on the car told the mine supervisor which miner got credit for the load. If the coal was too dirty, or rather, if it had too much rock mixed in it, the check also let the supervisor know who to yell at. Larger coal operators along Georges Creek, particularly the Consolidated Coal Company, would dock miners' paychecks for dirty loads.

Young boys would usually work at the sorting area as their first mining job. They sorted the coal from the rock and slate even more than the miners had done before shoveling it into the coal cars. The Wolf Den Mine's coal seam was more than seven-feet wide, but less than six feet of the seam was coal. The rest was slate, bone and rock that needed to be removed so that the coal would be clean. Basically, a customer paid for coal not the rest of the stuff around the coal. If there wasn't enough coal in each ton, the coal wouldn't burn efficiently and the customer wouldn't be too happy. So the coal needed to be "cleaned" before it was loaded into the tipple and then railroad cars.

Cleaned coal was weighed at the tipple and then dumped into a chute, which dumped the coal into railroad cars. The railroad cars sat on a short spur of the Western Maryland Railway that connected to the main line in Dodson.

Miners in Dodson relaxing. Photo courtesy of the Western Maryland's Historical Library.

When the miners walked out of the mine at noon for their lunch break, they squinted for most of the time even if it was a cloudy day. Some of the miners would head home for lunch, but they were usually the single men. Married miners had wives who hated to see them tracking more coal dust than necessary into their homes. So the married miners carried their lunches with them in their lunch pails, which looked like metal buckets with lids. Lunch typically included tea or coffee along with a sandwich and dessert.

Jack Ayers, a miner from nearby Barton in Allegany County, once said that miners always ate their desserts first in case of an accident. "At least we always had our dessert," he said.

The miners lounged in the shade of the mountain, talking and eating. You could see an interesting mix of men then. You had young men just out of high school who were still not used to spending so much time in the dark. They moved around a lot and stood and stretched as if trying to grab the sun. Old miners who had spent much of their adult lives underground had poor vision after so many years in near darkness. They squinted in the light and just enjoyed not having to work for an hour.

Then there were men who had been crippled for one reason or

another, but still worked in the mines. They might be missing an arm or leg or they might even be like John Schooler. He was a miner in Nethken who had been crippled by polio. He was only able to walk around on crutches. John worked in the mines as a digger despite his handicap. He had two sets of crutches for his work. He had his regular crutches for getting around, but he also a pair that was half as long. He used these shorter crutches to pull himself through the tunnels to get from the layoff to the face. When he needed to push a coal car into place, he would grab hold of the track to anchor himself and then use his shoulder to push the coal car.

George Brady and his future father-in-law, Fitzhugh Burrell, were working outside of a coal mine they called Pee Wee near the tipple.

Fitzhugh told George, "I'm going inside to help the guy on the cutting machine."

George waved to him and Fitzhugh started walking into the mine. The cutting machine drilled under the coal and someone needed to be at the face after the machine passed shoveling the coal out. That's where Fitzhugh was when a large rock fell on his back.

He was on his hands and knees at the time and the rock pressed him to the ground. It was so heavy that it knocked the breath right out of him, and he couldn't draw another one. The last thing he heard before he passed out was footsteps of the other miner running off to get help.

Fitzhugh kept thinking, *Please don't leave me. Come get this rock off me because I'm not dead.*

When the miners returned, they pulled the rock off of Fitzhugh. They thought he was dead because he was unconscious and not breathing. His heart was beating, though, so a miner gave Fitzhugh chest compressions until he was breathing again.

The miners carried Fitzhugh out of the mine, loaded him into a car. George went with him to the hospital. Fitzhugh was suffering from a broken back, leg, ribs and ankle. When he regained consciousness and heard what had happened to him, Fitzhugh told the doctor that he wanted a cast that would still allow him to walk.

"Mr. Burrell, you may not have to worry about a cast. You may not ever be able to walk again," the doctor told him.

However, the doctor fitted him with a limited cast and George drove him home that evening. Where most people would have wanted to rest, Fitzhugh was doing exercises before he went to bed that night.

He eventually recovered and went back to work in the mines

shoveling coal as if he hadn't been injured.

Coal miners are tough men who refuse to be stopped when they want to do something.

Just about the time the diggers would get their backs unkinked during their lunch break, they would have to return to the mine and work for hours more bent over. When the diggers showed themselves again at six p.m., they would be as black as coal tar. Some of the miners were so tired that they simply trudged home carrying a sack of coal if it was needed. A monthly coal charge was deducted from their wages for coal so they made sure to use it. Again, those were the single miners. The married miners would at least wash their faces, necks and hands with Lava soap before heading home.

Lava soap, which contained pumice, was the only thing that could effectively cut through the coal dust and dirt from a day in the mine. The pumice made it a bit rough, but that is what helped it work. Even using Lava soap to clean up took some scrubbing and the bath would have to be refilled a couple of times because the water would get so black from dirt and coal. A bath in those days meant sitting in the kitchen in a number three wash tub filled with water heated on the stove. Water tended to be either too hot or lukewarm. That's why most miners only bathed once a week. They figured why go to all that trouble each day only to be just as dirty the next day.

Their dirty clothes would be washed the next day. Since most miners had only a few changes of clothes, this meant that each set of work clothes was washed multiple times each week.

Once the miners dressed in clean clothes, they would eat dinner then relax for an hour or two before bedding down and starting the cycle all over the next day.

As hard as the work was, it was a job that the miners thought would always be there. Things started to shift when demand for coal in Maryland decreased sharply after 1920.

People wanted gasoline and fuel oil for heat and power not coal. Less demand meant less money paid for the coal. Coal company profits dropped and so the pressure to cut wages or employees or both increased.

Up until that time, the United Mine Workers union had been finding it hard to get a foot hold in Maryland because work conditions were good, at least as far as mining work went. Those men who did join the union had to do so in secret, meeting in the woods far from the coal company's eyes and ears. Being a union member could cost a Maryland miner his job in the early years of the twentieth century.

The UMW called a national strike on April 1, 1922. Though most of Maryland's miners were non-union, they walked out in support, including those miners in Shallmar. The miners were earning between $6.40 and $7.20 a day. The union was pushing for $7.50 a day.

In Maryland, the UMW used the strike to help push its unionization effort for Maryland coal mines. Previous attempts to unionize Maryland miners had been made in 1879, 1882, 1886, 1894 and 1900.

Though Maryland diggers had issues with whatever mining company was running the mine, at times they also had reason to doubt the union's commitment to them. During the 1882 strike, the National Knights of Labor failed to financially support striking miners, which weakened the miners' ability to continue striking. County miners remembered this when the 1894 strike call came and some miners refused to strike because they feared the union wouldn't financially back them.

When the UMW national strike ended on August 15, the Garrett County miners stayed out in an effort to win union recognition. The UMW supported the strikers with money and a food commissary. The mining companies, for their part, brought in strikebreakers from Cleveland, Pittsburgh and West Virginia. They were armed with automatic weapons and even machine guns.

Some of the conflicts were minor like the time that a young Kenny Bray built a snowman in his front yard that was holding a sign with the word "SCAB" written on it in large letters so that it could be easily read from his neighbor's house. The neighbor was a miner who had crossed the picket line and wasn't too happy to be called the slur for a non-striker.

"He cursed me and ran me into the house," Kenny said. "My mother came out and ran him into his house."

You don't cross a coal miner's wife, especially one who is also a mother. It's like trying to hold a greased rattlesnake and just as deadly.

In one instance, union miners' wives planned to jump the non-union miners in one neighborhood. The wives were armed with clubs and rotten eggs. One woman even had a bush with plenty of thorns on it that she planned on using as a painful switch. The women hid behind a building lying in wait for the miners.

The first miner came walking along. Kenny said, "The women came out and began to beat him, one woman hit him across the head with the bush knocking his hat off. Another woman began to slash him across the rump with a pick handle, when he stooped to pick up his hat. They hit him with only one egg. They were saving the other eggs for the next man."

The first man ran off to his house and the women laid in wait for the next man. However, the first miner returned to warn the unsuspecting miner. The first miner also had the women arrested for attacking him. During the trial, the miner showed the justice of the peace where he had been hit with the bush. One woman in the audience jumped up and told the miner to drop his pants and show where she had hit him with the pick handle.

Another time, union miners blocked the one-lane bridge across the North Branch Potomac between Kitzmiller and Blaine. They drove two cars toward each other from opposite ends of the bridge. When the cars met near the middle, no one could drive around them. The drivers got out of their cars and began yelling at each other for the other one to back up.

It was all a ruse. The performance had been staged to block the bridge at just the right time. A truck from one of the mines came rumbling down the road carrying supplies for the mine. It couldn't cross the blocked bridge, though. So the driver stopped the truck.

That was just what the union miners wanted. A group of them rushed from their hiding places and placed boards with nails sticking out of them under the truck's tires so that the truck wouldn't be able to move without blowing all of its tires.

The driver was prepared for union violence. He pulled a shotgun off the rack behind his head and stuck it out the window and warned the miners to clear off.

That gave the miners pause since none of them wanting to be shot. A woman from Shallmar ran up to the truck and jumped on the running board. She grabbed the barrel of the shotgun and placed it against her breast. Now she was a big, buxom lady so she had plenty of breast to cover both barrels of that shotgun and then some.

"If you want to shoot someone, shoot me, you son of a bitch," the woman yelled.

The driver, well, he couldn't bring himself to shoot a woman, especially an unarmed one, though he might have been thinking about it. The important thing for the miners was that the driver hesitated. That gave the miners time to get the shotgun from him. It went flying into the river.

The miners then beat holy hell out of the driver. He didn't die, but he spent some time in the hospital recovering from the beating he took.

Things just got worse. Striking miners in Western Maryland would shoot at non-union miners who were still working. In the extreme instances, people were even murdered.

The UMW called off the nearly 20-month strike in November 1923 without unionizing the mines. What did happen was many miners lost their jobs not only because the mining companies were careful who they rehired, but because the strike crippled the mining industry in the area.

It wouldn't be for another decade or so that the UMW gained a foothold in the state and it was because of Shallmar. During another strike at that time, women were marching through the streets of the coal mining towns, beating their pots and pans to raise enough racket that you would have thought it would cause a landslide. The men stayed home and didn't go to the mines.

The coal companies didn't take these actions lying down. Once again, they brought in Coal and Iron Police, which was what the coal company officials called strikebreakers, to try and work the mine and break the strike.

In Shallmar, the striking coal miners began picketing the company store and blockading the road to the mine. It was a tense time. People feared that it might turn into another strike like the 1922 strike. The coal companies might have held out, except that Shallmar Coal Company was the weak link. Wilbur Marshall came down from New York on one of his infrequent trips and signed a contract with the union. He was the first of the Maryland mines to do so and the rest soon followed his lead.

Wolf Den Coal Company failed in 1927, but it was reformed as a leaner Shallmar Mining Corporation. Otherwise, it was the same company and the same officers.

A pay slip from Kenny Bray that shows how much
a miner made and what was paid from his wages.
Courtesy of George Brady.

A leaner company meant costs were cut. Lower wages were ne-
gotiated. Paint began to flake off the houses and the hedges began
looking a little wild. Some miners stepped in and took care of their
properties but not everyone. Still, Shallmar was a nice place to live.

Then the Great Depression sucked under Shallmar Mining Corporation like a swimmer caught in a riptide. The company declared bankruptcy in 1938 and was placed into receivership with Howard Marshall as one of the receivers.

Howard was Wilbur Marshall's nephew and his uncle had been grooming him for years to run the company. Howard had come to Shallmar as an assistant mine foreman just after the Wolf Den Coal Company had opened.

Despite what he would tell you, Howard didn't know anything about mining. Right before the Wolf Den Coal Company had opened, Howard had been working as a "trainman" in New York. Now to be an assistant coal mine foreman, he eventually did have to learn a minimum amount about the business in order to pass a written and oral exam that Maryland began requiring in 1923.

Howard was a young man in his twenties who had an engaging personality and he liked people. These were qualities his uncle was hoping to refine for his own benefit when he hired his nephew on at the coal mine.

Despite living and working among coal miners, Howard never quite understood the mining life. He was one of the few people in Shallmar who had a car and he liked for everyone to know it. To make sure they did, he liked to drive it backwards through town raising a racket as he did. That way he not only got to show off his car, but he got to show what a good driver he was. He also made sure that his cars were never more than a year or two old. Howard had some family money so he wasn't dependent on his mining wages alone.

After a few years as assistant mine foreman and mine foreman, Howard became the mine superintendent of the Wolf Den Mine in 1923. He continued to learn as superintendent, but now he learned more about what his uncle did in New York to gain coal contracts than the best ways to mine coal.

When the Shallmar Mining Corporation declared bankruptcy in 1938, he and W. G. Hobbs were appointed the receivers for the company. It was their job to try and keep the company solvent while the bankruptcy process worked itself out.

That wasn't such an easy job to do. The Wolf Den Mine had gone from averaging over 100,000 tons of coal a year to 77,000 tons a year during the Great Depression years and its worst year of just 25,000 tons came during the time the company was in receivership.

Glosser and Sons Company out of Johnstown, Pennsylvania, purchased Shallmar and the Wolf Den Mine at a bargain price during this time. However, Howard wasn't done with the coal business.

He put together a new company called the Wolf Den Coal Corporation and convinced his brother-in-law, Jesse Walker, to become the mine foreman. Jesse had been working as the foreman at the Hamill Coal and Coke Company in Blaine, West Virginia, which was across the North Branch Potomac from Kitzmiller. They went into business together and leased back 3,000 acres from Glosser and Sons Company.

Though Howard was listed as the president and superintendent of the new company, he embraced his role as president more so than superintendent. He left Jesse to handle most of the duties of the superintendent while he spent much of his time in New York City trying to win business for the Wolf Den Coal Corporation.

That suited Jesse fine. He quickly realized that Howard was a much better agent for the company than a mine superintendent.

"Howard doesn't know anything about a mine," Jesse said when asked.

When World War II broke out, boom times returned to the coal mines because the Navy's ships needed coal to power them. The miners should have known that the good times wouldn't last or maybe they did and didn't care. Their work forced them to live in the present because tomorrow might never come.

For instance, Dottie Crouse had seven children plus her husband all living in a small bungalow in Shallmar during the war, but rather than save her husband's pay so that they could afford one of the larger homes on the main street, she rarely left the company store with any cash or flickers.

As soon as the daily sheets came down from the mine with the daily totals of each miner's tonnage, she was in the store getting a coupon slip for the pay owed. With coupon in hand, she then spent the next half hour or so buying things from company store whether she needed it or not. If after buying what she needed, she still had money left over, she would buy candy for her kids.

Victory in Europe and Japan spelled trouble for the coal fields, though. The boom times ended and coal prices fell once again. Small mines like Shallmar's were the first ones to feel the effect of those falling prices because they operated on a smaller profit margin than larger coal mines. Hours were reduced. Periodic shut downs began and then grew in length. Longer periods of time went by between equipment maintenance.

No one called Shallmar a nice place to live any longer, but then, it wasn't something said of any coal town. The U.S. in general was enjoying a post-war economic boom. Americans had saved and sacrificed their normal purchases in order to free up resources to be used for the G.I.s fighting the war. Once the war ended, all of that consumer demand that had been held in check during the war was released and Americans began to spend, spend, spend. They bought houses. They bought cars. The country's gross national product grew by around 50 percent during the 1940s.

It didn't reach the small coal towns in Western Maryland, though.

Oh, the families were happy when their soldiers returned alive. These veterans were not only lucky to be alive, but they also had the means to leave Shallmar and many of them did. The G.I. Bill paid for veterans to get a college education, provided them with a year of unemployment compensation and offered low-interest loans for home purchases.

The world seemed to pass by those who remained in town. Maybe it was a good thing that many of them didn't have cars. They couldn't drive to Oakland, Keyser or Cumberland and see what their lives might have been like. What they read in newspapers or heard on the radio seemed a world away. Their world was Shallmar and the mile or two around it, but it was a world that was slowly disappearing and threatened to take them with it.

5

Shallmar's Principal

Following Paul's discharge from the U.S. Army, Paul and Molly Andrick returned to Philippi and reunited their family. It had been months since they had seen Jerry and they were missing him.

Paul was now free to return to teaching, but he didn't have a job waiting for him at Bayard High School. His former job had been filled when it looked like Paul's stint in the army might last a year or more. No other promising teaching jobs loomed on the horizon either.

When Paul had been teaching, it hadn't taken him long to see where education needs weren't being met in the classroom. Textbooks were old and worn with missing pages and the scribbles of many of the previous students who had used them. The classrooms lacked basic supplies, such as chalk, paper and microscopes. He saw lots of inefficiencies in the bureaucracy of the school system. Having to deal with all of it irked Paul.

He wasn't someone who would sit around and see "what happened." If he had done that on the farm, his chores would have piled up beyond his ability to do them and would have affected his family. In college, he would have starved if he hadn't worked to earn extra money. His years teaching had given him some ideas on how things could be improved. He never had the authority to make the changes so nothing happened, but what if he did have the ability to make the changes?

Paul and Molly had talked about their future plans on the train ride north. As a disabled veteran, Paul not only qualified for the G.I. Bill, he would receive a small disability pension from the U.S. government. He now had the means to earn his master's degree. That, plus the fact that he was a veteran, would help him land a job. He could earn his degree and become a school principal or a supervisor.

Paul applied and was accepted at West Virginia University in Morgantown, West Virginia. There was no question that he would be. He had excellent grades, was a veteran and a West Virginia resident. In other words, he was just the type of student that WVU wanted.

The Andricks moved to Morgantown in the summer of 1945 and Paul prepared to become a student again. A contemporary of Paul's at the university would be a young comic who was also a fellow veteran and West Virginian named Don Knotts. Knotts would go on to play his Emmy-Award-winning role as the bumbling, but lovable, Deputy Barney Fife on "The Andy Griffith Show," which painted an idyllic picture of small-town life. Paul would go on to actually live in a small town much different than Mayberry.

Paul and Molly bought a house near campus and rented part of it to other students. The rent they received provided them with some additional income to make ends meet. It was a big help to Paul who had little time to work an extra job between trying to finish his studies as quickly as he could and helping raise his family.

He graduated in 1947 with a master's degree in administration. His first administrator position was as a teaching principal at Red House School, in Red House, Maryland. The most-recent schoolhouse there had been built in 1936. It was a two-room brick building with a basement and auditorium. Once it opened, the Garrett County Board of Education had consolidated smaller schools into it. The Gaver and Red Oak schools were consolidated in 1936. Corunna School consolidated in 1940 and Gorman School consolidated into Red House in 1944. The result was that Red House School was packed with kids. The auditorium, which was in the basement, was converted into classrooms to handle the additional students. It wouldn't be until 1956 that a two-room addition was built onto the school to alleviate the overcrowding.

Paul replaced Hildred Mulvey who had been the teaching principal at the school since 1943. He oversaw fellow teachers, Anna Sue Harvey and Alta H. Duling, as well as taught at the school. His administrative responsibilities brought with it an increase in pay, which he appreciated.

Still, it wasn't an easy decision to take the job. Although Red House in Garrett County was relatively close to West Virginia, it was too far away for a daily commute, especially over mountain roads that could be treacherous in the winter. This meant that Molly wouldn't be able to teach unless she could complete her work for her bachelor's degree.

During the 1947-1948 school year, the Andricks rented the upstairs apartment of a house in Gorman, Maryland. It was a small town that grew up around the Hoffman and Company Tannery and close to Red House.

His first experience as an administrator went fine. Paul instituted his policies at the school without any staff disagreements. When a chance came to move to a bigger school at Shallmar the following year, he took it.

By the time Paul arrived in town to be the principal and one of two teachers at the Shallmar School, the town had just about hit bottom. The Wolf Den Mine operated only thirty-six days in 1948. For miners who only got paid when they worked, it was like taking a pay cut of about eighty-five percent and even that pay came late and was paid out in company scrip. The miners fell behind in their bills and this caused them to be unable to leave town even if they had had the opportunity. They simply couldn't afford to live anywhere else because the mining company didn't press them for rent when the mine shut down.

Folks in town put on a good face to welcome the Andricks. The miners who hung around the company store talking stopped Paul on the street to say, "Hello." Though he wasn't one of them, he was responsible for teaching their children and the miners knew that an education was what their kids needed if they were going to get out of coal country. The miners' wives stopped by the house to talk for a few minutes with Martha. Six-year-old Jerry had the easiest time of it as he made new friends among his classmates.

But this was just the side of folks that Paul, his family and most everyone else were allowed to see. It wasn't that the residents were trying to mislead anyone. It was the way that the miners dealt with hard time. It's like if you talk today with a senior citizen and ask him, "What was it like growing up poor?" more than likely his reply will be, "We didn't know we were poor because everyone around us was the same way." That's the way it was in Shallmar then. Folks were all suffering with no work, no credit, no savings and no hope for any of changing in the near future. But why should they complain? Everyone around them was in the same situation and no one knew how to fix it.

But Paul came to town because he had a job waiting that paid good money. He didn't need to buy the overpriced goods at the company store because he had a car and could drive an hour or so north or south and be in Oakland, the Garrett County seat, or Keyser, the Mineral County, West Virginia, seat. Actually, he preferred not to have to buy from the company store since he didn't like to see black thumbprints on his cheese, which seemed to happen every time Baxter Kimble was doing the cutting since the man was notorious for not washing his hands. Plus, while the Wolf Den Coal Corporation expected its workers to buy goods from the company store and somehow knew if they didn't, the Garrett County Board of Education employed Paul so he would shop where he wanted.

He, more than anyone else, should have seen what was happening, but he didn't not until it was almost too late.

Students in grades 1-6 for the 1951-52 school year on the steps of Shallmar School. Front row (l to r): Susan Watson, Cathy Hartman, unidentified boy, Penny Fazenbaker, Floyd (Butch) Brady, Catherine Burgess, Charlotte Crouse. Second row (l to r): Thomas Richard Watson, Nancy Jo Swansboro, Marion Fazenbaker, unidentified girl, Bobby Hanlin, Ross Dale Lyons, unidentified girl. Third row (l to r): William Robert Tasker, Bobby Hartman, Jerry Paul Andrick, Mary Louise Burgess, Billy Pyles, Billy Crouse. Fourth row (l to r): Billy Hartman, unidentified girl, Robert Lyons, Junior Males, Danny Melavic, Jerry Melavic. Photo courtesy of Jerry Andrick.

6

A Kid's Life

Shallmar was among Jerry Andrick's earliest memories. For a six-year-old boy, Shallmar wasn't an economically depressed coal town, far from a city and verging on ruin. Those types of thoughts didn't enter into Jerry's mind. For him, Shallmar was a place of fun, adventure and mystery as only a young boy can see the world.

He had lived in Morgantown, a town nearly 100 times larger than Shallmar, though he couldn't really remember it. He'd also been to Keyser, West Virginia, to visit relatives. That city was also much larger than Shallmar. Keyser was smaller than Morgantown, but he could remember it better than Morgantown. Both Keyser and Morgantown were noisy, busy places where his parents watched him much closer than they did in Shallmar.

In Shallmar, he could run and be wild and there weren't that many people to disturb. At night, he could stand outside and see every star in the night sky. The town had woods to explore, a river to swim in, a mountain to climb and a hole where very few people went into and when they came out, they were changed.

Take his neighbor, Mr. Fuzzy Burgess, who was neither hairy or an elected official. Arthur Burgess was a nice man with a funny nickname who lived with his wife and two daughters in the house next to the Andricks. Jerry only knew him by his nickname and thought that it was his real name. Jerry thought that Fuzzy's parents had played a crueler joke on their son than Jerry's parents had played on him by naming him J. Paul Andrick without being named after his father. It was odd having two J. Paul Andricks in the same house.

Every once in a while, Mr. Burgess would walk with some other men into the hole in the mountain. He'd be gone for most of the day and when he returned, he would be as black as a piece of licorice.

Only later did Jerry learn that Mr. Burgess was a coal miner who was covered with a layer of coal dust from working in the Wolf Den Mine. However, by the time Andricks moved to Shallmar, the mine was barely operating so Jerry didn't have many chances to see the miners tramping off to work in the early morning.

His mother would use the sight of Mr. Burgess returning home as a teaching moment to tell him about the importance of a bath and keeping clean. Jerry really didn't mind taking a bath. It just seemed like his mother always wanted him to take his bath when he was busy playing with his toys or listening to one of his favorite shows on the radio when they could get reception.

The Andricks moved to Shallmar in the summer of 1948. That left Jerry free to spend the long summer days making friends before he started the first grade in the fall.

One of his first friends was Bill Crouse who was part of a large family that lived in town. Bill had three older brothers, an older sister and a younger sister. Jerry was an only child. He had a bedroom all to himself while Bill slept on a pallet on the floor of the living room with his brothers. His parents shared one bedroom and his sisters shared a bed in the other bedroom of their house.

Jerry couldn't understand why Bill seemed so frustrated with his brothers. Jerry wouldn't have minded having a friend to play with all of the time. Of course, Bill's older brothers and sister did seem to try and boss him around a lot.

Jerry Andrick

Having been born and raised in Shallmar, Bill could show him everything a kid needed to know to get by in town.

The best deal in the area for a kid was at the Maryland Theater in Kitzmiller. For twenty cents, you could watch movies virtually all day. Your ticket allowed you to watch a movie serial, a cartoon and a movie. For an additional nickel, you could buy yourself some candy or popcorn to munch on while you watched Hopalong Cassidy and the Cisco Kid on the big screen.

Since money was hard to come by, kids had learned to scavenge soda bottles from around the area. The company store would return the deposit for every bottle you turned in. At a penny a bottle, twenty bottles would be enough for you to buy a ticket on Saturday afternoon. The problem was that with all the kids looking for bottles, they got harder and harder to find.

Not all of Bill's lessons were so productive, though. Bill showed Jerry how to climb to the top of the gob pile without slipping. You had to be careful climbing it because although it looked solid, it wasn't. It was the pile of stone and dirt; waste material produced from mining the coal in the Wolf Den Mine. If the area had had a lot of rain, then there was always a chance that the heap could give way as if there was a mud slide. It had happened before with the gob pile at Shallmar and it was only a matter of time before it happened again. You didn't want to be standing on it when it suddenly started moving.

The first time Jerry and Bill climbed to the top of the pile, Jerry looked around and felt that he was standing on top of the world. "You can see forever," he said.

Bill wasn't interested in seeing how far down the river he could look. He was studying the gob pile and where it bottomed out.

"Follow me," he told Jerry.

The boys took off running down the mountain of debris. They accelerated quickly on the steep slope. Jerry felt like he was flying because it seemed like his feet were only touching the pile every twenty feet or so.

The boys screamed in a mix of delight and fear. For Jerry, it started out as mostly delight, but then he saw the large rocks at the bottom of the slope approaching too quickly. They hadn't looked so big from the top of the pile and he didn't have time to stop.

Then Jerry watched as Bill dropped to his backside on the pile and slid the last bit of the slope stopping before he hit the rocks. Jerry quickly did the same, feeling the small rocks in the dirt grab at the back side of his jeans, slowing him down.

When they both had halted, Jerry and Bill started laughing. Leave hopscotch in Shallmar's street to the girls. The boys had the gob pile.

Playing on the gob pile was something that all the Shallmar boys enjoyed doing and despite the apparent danger, only one boy was ever seriously hurt while Jerry lived in town. He fell the wrong way and broke his arm. Injuries were usually just cuts or scrapes and the occasional whipping from a mom who was mad that her son had ripped a pair of his good pants. Of course, when your butt was numb from sliding on small rocks for half a day, a whipping didn't really hurt.

Eventually Jerry also learned how to summer sled, which was sliding down the gob pile on an old board. The boys would get warped boards from over where the old tipple had burned and carry it with them to the top of the gob pile. Then they would slide down the pile on the board, making sure to roll off the board near the bottom so they didn't crash into the rocks. It was even more fun than running down the gob pile.

It was all great fun, but it left the boys looking nearly as black with coal dust as the miners after a day's work. Of course, the payment for such fun was that Jerry's mother would lecture him about cleanliness and make him take a bath. He wasn't sure, but his mom may have even scrubbed a little harder than necessary to get him clean. Now that could hurt, especially if she was scrubbing him with a brush.

Since it was summer, the boys would spend a lot of time cooling off in the water. Besides, it was a better way to wash off a coating of coal dust after a few trips down the gob pile. The best place to swim was in Harrison. The boys would walk across the swinging bridge, which was made from steel cables and wooden slats strung across the North Branch Potomac and hurry over to splash in Abram Creek. The water there wasn't orange like the water in the North Branch Potomac. It flowed clear and cool and the swimming hole had a flat rock bottom so it didn't bother your feet to walk around in the three feet of water. Of course, that didn't stop the boys from leaping in the river for a dip if they didn't feel like hiking to Harrison. Then they would have a layer of slime on top of the coal dust, but at least they would be cool.

At the end of a hard day of playing, they might stop by one of the kids' houses in town for a snack of home-baked bread slathered with apple butter or jelly. Though most people enjoy fresh-baked bread, the older kids who attended Kitzmiller School were actually embarrassed to be seen eating homemade bread. They said it wasn't cool. Most of the kids who went to Kitzmiller School ate store-bought bread so the kids from Shallmar stood out.

Besides being more food conscious than their younger brothers and sisters, the older kids in town were a bit more mischievous in their free time. At least the boys were. According to them, you could only play ball or swim for so long before you got bored.

Once, a group of boys in town noticed that when a young miner would come walking down the street, he would whistle a particular tune when he walked by a house where an attractive girl lived. Hearing the tune, the girl would come outside and go walking off with the young man. The young man couldn't go to the girl's door himself because her parents didn't approve of him.

Sometimes, the boys in town would stand on the street whistling the tune just to get the young woman to rush outside. When she saw it wasn't her beau, she would shout at the boys who would run off enjoying a good laugh.

Another way they tormented this couple was to follow them at a discreet distance. The couple would generally cross the swinging bridge and go somewhere in Harrison since they didn't have to worry about being seen there. Once the couple had crossed the bridge and walked on, the boys would sit on the Maryland side of the bridge and sing songs until late in the night. The couple would return, but they wouldn't be able to cross back to Shallmar because they didn't want to be seen. It surely made for a lot of explanations in the girl's house when she came back so late.

The boys also discovered another couple who wanted to keep their rendezvouses a secret. This time it was because the woman was married. A miner in town would stop in to see his girlfriend when her husband was working evenings at the mine. When the husband came home, the wash tub hanging on the wall out back would rattle as he stepped on the back porch and moved toward the back door. Miners always came in the back door when returning home from work so as not to track coal dust through the house. Anyway, the rattling tub was the signal to the miner inside who was making time with another man's wife. He would bolt out the front door and casually walk down the street.

Once the Shallmar boys realized this, they snuck up to the house one night and rattled the tub. Then they ran around to the front of the house and watched the miner scurry out. When no one entered the house, the wife called the miner back inside. The boys waited a few minutes so that the miner could settle down and start making time with his girlfriend. Then they crept back to porch and rattled the tub again. The miner bolted out of the house again. The boys covered their mouths to hold in their laughter as the wife called her boyfriend back. The boys snuck back and rattled the tub a third time.

This time, the miner didn't run out when the boys ran to the front corner of the house. The boys waited.

"You guys are going to cause me to break my damn leg!" The miner had come out the back door and snuck up behind the boys.

This time it was the boys who bolted up the street, laughing as they ran.

Most boys in coal country grew up knowing that someday they would become coal miners. They would drape blankets across chairs and tables forming their own mine tunnels to play in. Their first jobs would be working at a mining company's picking table and eventually they would follow their father into the mine to be trained as a digger.

They also learned how to work with explosives. It was a skill they weren't supposed to learn until they were actually diggers, but boys could be clever and inventive when there was mischief to be done.

One time, a teenager took an old carbide can outside behind the company store. It was quite a bit larger than the typical can since it had held 100 pounds of carbide. The boy and his friend scraped out remnants from a black powder can and added it to the carbide can, which also had a bit of carbide left in it. Then boys added some water, closed the can and ran.

The water reacted with the carbide, which started to burn and ignited the powder. The pressure exploded the can and flung the lid from the can a hundred yards away.

People inside the company store at the time thought there had been an explosion up at the mine. They ran outside and looked up at the mine with panicked eyes only to see the diggers at the mine looking back at them with the same expression of fear and confusion.

The first inkling that Jerry got about how poor his friends' families were was something that he didn't even recognize as such.

Jerry had his own bicycle and he liked to ride it on Shallmar's dirt road. He could ride it up near the entrance to the mine and then race down the hill, kicking up a cloud of dust behind him. If he had a friend with him, he liked to race down the hill to see who could reach the bottom first.

The trouble was he couldn't always find a friend to race with. Well, it wasn't that he couldn't find a friend, but he couldn't find a friend with a bike. While Jerry's bike was always available when he was, most of the children in Shallmar shared a bike with their siblings. The six Crouse children had one bike that they all had to use whether they were boys or girls. Jerry would let Bill use his bike sometimes when Bill couldn't get the family bike, but that didn't help if the two of them wanted to race their bikes against each other. There was nothing worse than having to walk when you wanted to pedal fast. It ranks up there with sitting in church on a beautiful summer day.

Jerry Andrick riding his bicycle down Shallmar's only street. Photo courtesy of Jerry Andrick.

Kids would play outside in the summer until the sun went down. Games like kick the can and hide and seek were very popular. But when the sun went down, the town was close to pitch black. Shallmar had no streetlights and the only light was the dim lantern light that shone through the windows of the houses in town. For Jerry, it was unnerving at first, coming from Morgantown where there had been streetlights at night.

Not everyone went inside when the sun went down. Jerry discovered that some of the men in town would play cards in the small growth of trees next to the company store. They sat on large rocks in the trees with kids standing behind them, holding lanterns so the men could see their cards and each other. Some weekends they would start playing cards on Saturday afternoon, then overnight and through

Sunday morning. They would only break up the game in order to go home and have time to get dressed in their baseball uniforms and get to the day's game.

The men offered Jerry a nickel a night, paid in scrip, to be one of the boys holding a lantern for them. He was excited because a nickel would mean a big handful of candy from the company store the next day. When he asked his mother if it would be all right for him to do it, she had two words for him: "Absolutely not!"

During the day, the men liked to play horseshoes and baseball if they weren't working. Most of the mining camps in Western Maryland had baseball teams that played each other. The men of Shallmar were also very serious about their horseshoes. They would measure the distance between the horseshoes and stake with a blade of grass to make sure no one eyeballed the distance wrong.

The new school year started that fall and Jerry and his new friends had class in the room for the younger students called the Little Room. "Little," in this case, meant the size of the students not the size of the room. The older students were in the Big Room where Jerry's father taught.

Even though he knew what to expect with school, it was still a big adjustment for him. He had to sit still at a desk for most of the day and not talk. For a young boy, it was quite a challenge, but he somehow made it through the school year. It helped that both of his parents were teachers and could help him with his school work.

Jerry looked forward to the summer of 1949, thinking that it would be just as fun as his first summer in Shallmar had been. He wanted to go swimming again and maybe even explore the coal mine. Kids weren't supposed to go into it, but when the mine was closed, no one was around to stop them. Some of the kids who had gone into it said it was pitch black and scary.

The summer started out that way, too.

In their explorations through the town and around the mine, Jerry and Bill found a dynamite cap and wanted to perform their own experiments with blowing something up. It was the greatest thing in the world to have until they realized that the cap was useless without the dynamite.

Jerry also learned to play baseball, though it wasn't in the typical way like adults played it. When he was bored, he would hit rocks into the North Branch Potomac using a broomstick. It may not have been like having a Little League coach, but it certainly improved his accuracy when he could hit a rock that was smaller than a baseball with a broomstick, which was thinner than a bat.

He wanted to go fishing, but he soon learned that the river had no fish in its orange water. That didn't stop determined boys from trying, though. Mine foreman Jesse Walker knew a pair of boys who tried fishing at the Dill Hole not too far from the Shallmar Company Store. One of the boys went to sit down on a rock and sat down on a snake. Now the snake didn't take too kindly to being squashed so he bit the boy on his butt.

The boy's fast-thinking friend pulled out his knife and had his friend drop his pants. He cut into his friend's butt around the bite mark and sucked out the poison. Anyone who is willing to suck poison out of your butt must be a true friend.

When the second boy was satisfied that he had gotten the poison out, he cut away the flesh around the bite marks and handed it to his friend who had decided that he would stand up while he fished that day.

"Here put this on your hook for bait," the second boy said as he handed his friend the small piece of flesh.

Now Jesse couldn't remember whether the first boy actually used a piece of himself for bait, but he did remember that both boys stayed to go fishing.

Most people chose to find a fishing hole somewhere other than the North Branch Potomac. Jerry liked to fish in Short Run on the western end of Shallmar. A boy with a fishing pole and some worms in a Prince Albert Tobacco can could have a lot of fun there.

Besides, not only was the North Branch Potomac off color, but it was full of trash. One of Jerry's chores each day was to take the trash can from the kitchen, walk down to the river with it, throw the trash into the river and rinse the can out.

Some of the other chores young boys might do during the day was splitting kindling for the morning fire, getting coal from the coal shed, helping their mothers in the garden and feeding any animals that their family might keep.

Dottie Crouse would have her son Bob wash dishes, which was a chore that he hated. So much so that when his mother wasn't looking, he would go outside and throw the glasses against the hill. He caught hell for doing it, but at least he didn't have to wash shattered glasses.

It was summer, a time for fun, and Jerry did have fun with his friends. For some of those kids, those days were quickly winding down. They were feeling the despair of their parents not only in the adult's attitudes, but physically, as dwindling savings vanished.

Billy Bray with a black snake at Shallmar. Photo courtesy of Jerry Andrick.

7

A Man and His Mine

If ever a person represented what was good about Western Maryland's small coal towns, it was an Alabaman named Jesse Walker. You'd be hard-pressed to find someone who knew Jesse and had a bad word to say about him. He was a man who didn't just live in his community; he was a part of it and worked to make it a better place to live.

Born in Blocton, Alabama, in 1896, Jesse was the sixth child in a family that would eventually have thirteen children. Blocton was a coal mining town near the center of the state and Jesse's father, William Walker was a coal miner. Blocton was still a fairly young town when Jesse was born. Truman Aldrich had built the town to service its mining company in 1882. The Cahaba Coal Mining Coal Mining Company became one of the largest coal companies in the south for a time.

Shortly after Jesse was born, his family moved further south in Alabama to Wilcox County, but around the turn of the century the Walkers moved to Kitzmiller. One of Jesse's uncles worked for the Hamill Coal and Coke Company and helped William get a job with the company. He quickly rose through the ranks to become a mine foreman.

Jesse attended the Kitzmiller School with his brothers and sisters until he was sixteen years old. At the beginning of his senior school year, his plans for graduation hit a snag. In the years before indoor housing had become common in the Western Maryland coal towns, the school had outhouses students used when they needed to relieve themselves. As it turned out, Jesse needed to do just that one day...really, really urgently. So he rushed out to the outhouse, but when he returned to the classroom, his teacher was angry with him for leaving and horsewhipped the young boy once the school day had ended.

When Jesse got home, his siblings had already told their parents what had happened to him. Jesse feared that he was in for additional punishment, but his father just looked at him and said nothing about it.

Jesse went to bed and his father woke him up early the next morning. Jesse was confused until he saw the lunch pail sitting next to the bed. His father took one of the round, oversized pails that were nearly as large as a bucket with him each day to the mine, but this one was to be Jesse's. He got dressed and walked with his father to the mine.

And that is how Jesse Walker became a coal miner.

The Hamill Coal and Coke Company's address was in Blaine, West Virginia, but two of its mines were in Kitzmiller. With the North Branch Potomac running through the mountains as it did, there wasn't always a lot of land consistently along either the northern or souther sides of the river. When the Baltimore and Ohio Railroad had been built, sometimes it was easier to run railroad tracks along the north side and other times it was easier to run them along the south side. Because of this, the railroad crossed the river several times. The coal companies needed a rail connection to ship their coal so if a spur line couldn't be run across the river a cable car system was used.

The Hamill Coal and Coke Company's Maryland mines had an 850-feet-long cable car system with a difference in elevation of 160 feet between ends was used to take the coal from the mines to the tipple on the West Virginia side.

Jesse's entry into the mine was coming of age tradition among coal miners as a father brought his son into the coal mine for the first time. Jesse knew some of what to expect. You couldn't live in a coal town and not know at least a little of what went on underground.

While Jesse's father probably didn't say much on that walk, the other miners, seeing a new face among them, kidded both of the Walkers. Jesse took the kidding in stride because that was his nature, but the experience of getting ready to mine coal for the first time both amazed and overwhelmed him.

His father helped him dress in work clothes, showed him how to operate his head lamp and where to get the supplies he would need for the day. Then they climbed into the coal car for the man trip into the mine. Instead of being able to watch the day brighten, Jesse watched it grow darker until the miners began turning on their carbide head lamps. The jarring nature of riding in coal car caused those lights to bob around casting odd shadows on the walls of the tunnel.

At the end of the heading, the coal car stopped and the men climbed out. Jesse followed his father through the tunnels to the room where they would be working for the day. The mines used a room and pillar design, though the rooms were larger than typically used in coal mines. The coal seams ranged in width from five to eight feet so diggers might find themselves working bent over one day and able to stand the next.

Jesse's father taught him how to shoot coal and then separate and shovel it into the coal car. It might seem a simple task, but a certain amount of skill was required to be able to find a rhythm that allowed the digger to throw each shovelful of coal with just the right amount of force to land in the car each time. At first, Jesse missed the car more often than not, as did most greenies in the mine.

William showed his son how to watch for rock falls, especially right after an explosion that might loosen rocks in the ceiling as well as the face.

Jesse was young and considered himself strong, but that first day in the mine drained him. He trudged back to the coal car in the main heading at the end of the day and slowly flexed his hands, which seemed frozen in the position used to hold the shovel.

He wasn't hungry at all when he got home and instead went to bed. When he woke up in morning, his body ached deep in his muscles. In fact, it hurt far worse and over more of his body than his teacher's horsewhipping had.

He got up the next day and followed his father into the mine once more and worked himself to exhaustion because that's what miners do.

The following fall, the teacher transferred to another school and Jesse's father sent him back to school to finish his senior year and graduate. When people asked why he had been gone for a year, Jesse always told them, "Father handled it."

Jesse was now a year older than the students in his class. It might have been uncomfortable for some teenagers, but Jesse didn't mind at all. He met a girl named Iva Bishop and he began going to the movies more often. It turns out that Iva played the piano for the silent movies that were being shone at the time at the Maryland Theater. Jesse probably couldn't have told you what the movies were about, but he could have described what Iva had been wearing that day.

Iva and Jesse got married after their graduation and Jesse started back at the Hamill Coal and Coke Company. He wasn't satisfied to stay a digger, though. He took college courses at night and began moving up the ranks at the coal company. Before too long, both he and his father were both foremen at the Hamill Mines.

Shallmar's houses from the back in the 1940s as the town started to enter its decline. Photo courtesy of Robert Hanlin.

When the Shallmar Coal Company went into receivership in 1938, Howard Marshall approached Jesse with a proposition. Howard's first marriage had failed in the 1920s because he had been fooling around on the side with Jesse's younger sister. After Howard's divorce, he had married Catherine Walker and Howard and Jesse became brothers-in-law. Now Howard wanted to become partners.

Howard might not have known anything about coal, but he knew how to schmooze and he could sell coal. He wanted Jesse to run the new company he was forming while he would secure the orders to keep the men working.

While Jesse would still only be listed as the mine foreman, his power with the new Wolf Den Coal Corporation would be more along the lines of a superintendent. Howard, of course, wanted to be listed as both the superintendent and the president of the company. As Jesse's niece put it, "If Howard was ever interviewed, he would have made himself governor of the state."

It was an opportunity for Jesse and he took it. Since Howard was planning on spending much of his time in New York, he offered Jesse and Iva the superintendent's house to live in. It was at the east end of the Shallmar's main road, not too far from the company store. Though the house was the largest in Shallmar, it wouldn't be considered showy in most cities. It was white clapboard with green blinds. It had a tower, which was an element of a style that had been more popular at the turn of the century. George Brady's mother was the Marshall's maid. According to her, there was a hidden area in the basement of the house where Howard Marshall had hid booze during Prohibition.

Iva took one look at it and said she didn't want to live in a monster house like that. They instead moved into one of the company's two-story houses. The house was larger than what the Walkers were used to and well built. It was plenty of house for them.

Howard and Catherine moved into the superintendent's house, though, as expected, they spent much of their time in New York City. Howard liked to show off and he was always inviting the Walkers to come to New York so he could show them the town.

Jesse gave in once and rode the bus up to New York City. He got off at the bus station and then found his way to the Waldorf Astoria Hotel, a fancy hotel even in those days, and just the type of place that Howard liked to stay when he was in the city.

Jesse walked up to the front desk with his suitcase and dog. The small dog was Jesse's constant companion. He had helped put his younger brother through college for which his brother was so grateful that he stole a puppy from the circus and sent it to Jesse as a gift.

Now Jesse hadn't known that the dog was stolen at the time. He simply got a call from his brother one day telling him to be at the train station when the evening train arrived. Jesse showed up and the porter handed him a box with a puppy in it. It turns out that the puppy's dam and sire had been trick dogs with the circus and the puppy inherited their intelligence. For instance, you could give the dog a note for someone at the school and tell him to take it to that person at the school and the dog would deliver the note.

Anyway, Jesse walked up to the front desk and asked for the room number for Howard Marshall. The desk clerk looked at the dog and said, "We don't allow dogs here, sir."

Now if the desk clerk thought that Jesse would choose the Waldorf Astoria over a mutt, he was quite mistaken. Jesse shrugged.

"Tell Mr. Marshall we were here," Jesse said. "We'll go stay someplace else."

The desk clerk probably didn't mind seeing Jesse go, but he certainly didn't want to anger one of his regular customers or risk having him take his business elsewhere.

Before Jesse had even reached the door, the clerk said, "Sir, I'm sure we can find a way to accommodate both you and your dog."

While Howard might have loved New York City, Jesse decided that he was much more comfortable back home in Shallmar.

When Howard was in Shallmar to act as the mine superintendent, things didn't necessarily run smoothly. The demand for coal was still falling and that meant the new coal company didn't need as many miners to fill its orders. The job of laying off miners fell to Howard.

At the beginning of January 1933, Howard laid off several miners, which meant that these men were also, in essence, losing their homes because they were no longer employees of the Wolf Den Mining Corporation. Late in the evening of January ninth, an unknown assailant shot Howard. He was admitted to Memorial Hospital in serious condition and remained there for several days until his condition had stabilized enough to be released. The shooter was never caught, but many people assumed it was one of the laid off miners.

Jesse took his responsibilities as the primary decision maker for how the mine ran seriously. He was at the mine every morning at five-thirty, a half an hour before the miners would arrive for the day. During that time, Jesse would walk every tunnel in the mine looking for any potential problems like loose rock or weak timbers. He prided himself decades later that in all the time he was the foreman for the Wolf Den Coal Corporation he didn't have any miners killed in fatal accidents and he only lost one pony.

It was one of the reasons that the miners respected him. He looked out for them and he didn't hate the union like most company men did. He realized that the UMW helped guarantee diggers good wages and health care.

However, when coal demand dropped off sharply after World War II, Howard simply couldn't get secure orders from people who didn't want or need coal. And no matter how much Jesse wanted to help his miners, he couldn't give them work that wasn't there for them to do. He could only hope that things would somehow change.

8

A Child Faints

At Shallmar School, Paul had been given responsibility for the education for sixty or so students who attended the two-room school depending on the year. The building had a third room, but it wasn't used for the school. It was the union hall.

When he walked over to the school for the first time in the summer of 1948, Paul found himself standing outside of it and just staring at the building. It was about what he had expected it to be. Nothing fancy like the multi-story brick buildings schools in Oakland or Cumberland. It was a wooden-frame, single-story building. It had three rooms that were part new and part the old Dodson School that had closed in 1930. That school had been moved to Shallmar, which was still considered the jewel of Western Maryland coal towns at the time, to become part of the Shallmar School.

Prior to getting its own school, children from Shallmar walked to Dodson to attend school there. However, the Dodson School was too small to house students from both towns. The original Dodson School built in 1905 had been a one-room school house. Within a few years, the number of students attending the school outgrew the available space and the Garrett County Board of Education had to rent the "theater" on the second floor of Dodson's community building to use for students in grades four through seven.

Once Shallmar had been built, the Dodson School had officially become the Dodson-Shallmar School and it stayed that way until 1930 despite the fact that Dodson's student enrollment was falling as that coal patch died. During this time, school board minutes noted, "there has been no school building in Shallmar. The children have been housed most anywhere in deserted Dodson; for instance, in almost any type of unsatisfactory buildings which were practically donated."

It was only when Dodson had almost no residents and it buildings practically falling down that the board of education had relented and given Shallmar its own school.

So why did Paul find himself simply staring at the school and not walking inside? Something was wrong, but he wasn't sure just what.

Finally he shrugged, giving up. The building must remind him of another one he knew.

He walked into the school. The two classrooms—the Big Room and the Little Room—were in the front of the building. Most of the desks, chairs and blackboards in the rooms had come from the old Dry Run School when it had closed. Each room also had a bulky cast-iron stove in it that burned coal, Shallmar's life blood. This was the school's heating system.

The hallway between the two classrooms ended at its intersection with another hall. The boys and girls bathrooms were off this hall as was the third room in the building, which ran across the back of the school and was the largest room in the building. It was the United Mine Workers Local Union hall.

Paul walked to a window of what was going to be his classroom and looked outside. He could see the mine entrance on the hillside. The coal company was shut down so there was no activity around it. It was just a hole in the mountain where one of the entrances to the mine was. Unfortunately for Shallmar, the shut downs were happening more often than they should if the miners hoped to make a decent living. The union could complain about the lack of work, but if the coal company didn't have any contracts for coal, it didn't need to be paying miners to dig it out of the mountain.

Paul turned away and left the school. When he did, he realized what had disturbed him earlier. Shallmar School had no playground equipment. There was no way for the young students to work off some of their boundless energy during recess.

That just wouldn't do.

Within a few days, Paul had called a meeting of all the fathers who had children attending Shallmar School. He walked them out to the empty school yard and asked them to imagine what it would look like with swings and a teeter totter. He told them that the school board budget for the school wouldn't cover the cost of playground equipment, nor could the coal company afford to cover the costs. It was up to them to help their children.

Money was in short supply in the region, but one business could provide some boards, another person might be able to supply rope and another nails. By asking for small donations from a lot of people, the group soon had all of the materials they needed.

They then gathered one day and built teeter totters and swings for the children to play on. Wives brought their husbands lunches to eat and children ran errands for their fathers and waited anxiously for the project to be completed. It was a long day of work for the men, but they were used to it and at least they got to work outside in the warmth of the sun this time. They also appreciated having something to do since there was so little work at the mine for them.

Children playing on the Shallmar School playground. Photo courtesy of Jerry Andrick.

As a final touch, Paul's group of volunteers painted the name of the school above the front doorway to show their pride in the school. It served the same purpose as the whitewashed stones on the river bank in front of the company store had once served. They were telling everyone who saw the playground and school that they believed in Shallmar and wanted to see it grow. This was their home and they took pride in it.

That had been a good day for Paul. It was a day that he thought he had made a difference in the lives of his students. He had heard their laughter and watched them play on the new equipment, and he had felt their joy.

With just three months of work in 1948, the Wolf Den Coal Corporation came into 1949 struggling to stay open. Howard Marshall and Jesse Walker were hoping that with of the State of Maryland's contracts for coal set to expire, Wolf Den Coal might be able to pick up some business. The State of Maryland purchased 70,000 tons of coal annually to heat its government buildings and it had a policy that every effort would be made to buy that coal from Maryland companies.

Children playing on the teeter totter built by parents for their children at the Shallmar School. Photo courtesy of Jerry Andrick.

The Wolf Den Mine was capable of supplying all of that coal need, though it wasn't likely to happen. Still, winning even ten percent of the state's business would help stabilize things somewhat and give Howard and Jesse more time to land another coal contract, perhaps one from the City of Baltimore, which also tried to buy its coal from Maryland companies.

Since the Wolf Den Coal Corporation was one of the largest mining companies in the state, it needed large contracts to justify its operations or a whole lot of small contracts. When the Garrett County Commissioners asked for bids for coal for the county courthouse in early 1949, you would think that all of the coal mines in the county, which were so desperate for work would submit bids. The commissioners received fifteen bids and they were from either individual coal miners who ran a single-person operation or very small companies. Most companies in the county wouldn't have been able to make any money off the contract for only a couple tons of coal.

The Wolf Den Mine passed its annual inspection in early March. Jesse Walker never doubted that it would. He took such care in making sure that his miners were safe that he could virtually tell you every noon and cranny of the tunnels. Federal mine inspector W. C. Ei-duke complimented Howard and Jesse on the mine's recent safety improvements. The mining company had improved ventilation in the mine and started doing rock dusting, which was applying rock dust into an area of a coal mine to reduce the risk of an explosion. The rock dust acted like rock in the mountain's coal. The more rock dust there was, the less efficiently the coal dust would burn.

Eiduke also noted that the mining company employed fifty-seven men who were capable of digging 326 tons a day out of the ground, more than enough to be able to fill nearly any coal contract. That wasn't even having the mine work at its full capacity or full employment.

However, the inspector also recommended that the mining company make some additional improvements. He wanted rock dusting to within eighty feet of the face in the mines dry rooms and entries to further reduce the risk of an explosion.

More air could be used in the work areas and an air lock large enough to contain an entire man trip on the main passageway. The mine doors needed to be kept closed except when in use, warnings needed to be posted at abandoned workings and records needed to be kept of the weekly checks for hazards. Jesse probably would have implemented all of the recommendations if he believed that they would make the Wolf Den Mine safer, but he also had to consider the cost. The expense to make these additional improvements would have been daunting to a coal company that didn't have any work.

The inspector's findings may have been the straw that broke the camel's back because the Wolf Den Coal Corporation shut down at the end of March not long after the inspection. Rumor had it that it was a permanent closure, but Howard and Jesse were still hunting for business. They submitted a bid on April 8 to supply Maryland coal, but when a state representative called the company with follow-up questions, he found out that the mine was closed. The company's bid for a coal contract was set aside without being considered.

What a conundrum! Customers wanted to buy coal from a company that could be relied on to be open, but the company needed orders to stay open.

So rumor began to look like the truth.

The town, the miners, their families and even the kids took the news of the mine closure hard. The joy left their smiles. They all wanted to know when the mine would reopen.

They weren't alone in that hope. In the fall of 1949, John L. Lewis called a strike right about the same time that coal demand was starting to rise. Even if the Wolf Den Coal Corporation had secured a coal contract, none of the miners could have worked on it, although all of them had seen their meager unemployment benefits end during the summer. Their families were in debt well over their heads to the company for their rent and to the company store for their food and now they had no income at all. They really did "owe their souls to the company store," as the song said, and they were wondering and worrying how they could keep food on their tables.

By the time that the national strike was settled during the first week of December, even miners who had regular employment were predicting a bleak Christmas for themselves. Though it was still weeks before Christmas, they only expected to have nine work days before the holiday and the money earned from that work would have to feed, house and clothe them. Any extra money, if there was any, would go towards helping them recover from the income lost during the strike.

The town got hit with more bad news after Thanksgiving (during which the Shallmar Miners hadn't found a lot to be thankful for) when the Garrett County School Advisory Board and the County Commissioners approved plans for a new county high school. Two plans had been introduced that involved redistricting and school closures to justify the construction of a modern high school south of Oakland. One of those plans, if pursued, would close Shallmar School and send all of the students there to Kitzmiller School.

Garrett County had nine two-teacher schools like Shallmar School. According to the annual Garrett County Board of Education report, $151.85 was spent per student to run Shallmar School, but only $1.30 or less than 1 percent of the school's budget was spent on upkeep. The school definitely needed some repairs and improvements, but the thinking at the county level was that it would be more efficient to close the school and send the students to the Kitzmiller School, which was in better shape and where the older students from Shallmar already attended.

The final decision was still a ways off, but it was apparent that Shallmar School, among others, was targeted for closure.

The future of his school wasn't something that Paul worried about, though. He had more-pressing concerns by the time that the report was released publicly that began to draw his attention.

He kept asking himself: How can someone watch another person die and not be affected? Had someone asked Paul that question the day before, he would have answered without hesitation that a person couldn't, at least not if that person had a heart.

Betty Mae Maule

That was yesterday. On this day, Paul knew that not only could it happen, it had happened to him. He was watching not just one person, but a room full of them, maybe even a town full of them, die and it hadn't affected him. He hadn't even recognized that it was happening.

He had realized all of this in a moment of clarity in the faces of Betty Mae Maule and her older sister Gladys. They were two young girls among the two dozen or so children in his class. The day progressed and Paul moved through his various lessons...English, writing, history, arithmetic. He didn't notice the change at first. He was too focused on covering the material that he had laid out for the day.

Paul was standing at the front of the classroom looking at Betty Mae, but not truly seeing her until that moment.

"Betty Mae, can you solve the equation?" he asked.

The ten-year-old girl stood up at her desk and chair. Paul watched the young girl's face go blank, and for a moment, he thought it was confusion that she might not know the answer to the math problem he had written on the blackboard. Then her eyes rolled back in her head and she fell to the floor.

Paul heard gasps of surprise, squeals of fear and shouts of "Betty!" Then there was the sudden scraping of desks on the wooden floor as some students moved to help her. Paul rushed forward and knelt beside her.

She was conscious. She still had a blank expression, but now it really did come from confusion.

"Betty, are you all right?" he asked.

Her eyes focused on him. "What happened?"

"You fell."

"I was standing and then I was on the floor."

"Does anything hurt?"

"No," she said in a weak voice. Now she looked a little scared.

Osborn Maule

Paul helped her sit up. He didn't see any blood on the floor. He felt the back of her head for lumps. He could feel one forming, but she didn't wince when he touched it. He helped her off the floor and into her chair.

Paul looked around and saw one of the Maule children. Betty had an older brother and sister in the Big Room, the class for grade four through six students. She had another sibling, Joseph, who was still in school, but he attended Kitzmiller High School. She was the baby of the family at least among her siblings.

"Osborn, has your sister been sick?" Paul asked.

Osborn shook his head. "No, Mr. Andrick. She's hungry. We're all hungry. She hasn't had anything to eat all day."

And that was the moment.

She's hungry.

76

Paul suddenly saw her...really saw her. Her face that had still seemed round with some baby fat at the beginning of the school year was now lean a little more than three months later. Her sunken eyes had dark circles around them. Her skin was so pale and thin that Paul could see her blood vessels pulsing beneath it and her bones pressing against it giving her face hard lines where there should be none.

We're all hungry.

Paul looked around and noticed that Betty Mae's sister, Gladys, hadn't stood up with everyone else. She was slumped at her desk. She was awake, but her face was pale and she had a vacant stare. Her eyes were magnified by her glasses, which sat crooked on her face.

He looked around at the faces of the nine, ten, eleven and twelve year olds in his class. Nearly all of them had the same look.

We're all hungry.

"Osborn, when did your sisters last eat?"

The fourteen-year-old thought for a moment. "Last night. We had a couple of apples at supper."

"Two apples were all Betty Mae has eaten since yesterday?" Paul asked, surprised.

Osborn shook his head. "Not just her. Three apples is all the family had to eat."

Paul looked between Osborn, Betty Mae and Gladys. There were eight people in the Maule family. That meant they had each only eaten half an apple or less since yesterday. It was past lunchtime now.

He knew that Shallmar's residents were impoverished with the coal company shut down. That had been more than eight months ago. With no work since then, all of the miners had to be in a bad way. How many of the school children had he seen come to school without shoes or with threadbare clothing?

How desensitized had he become to miss this? Paul had grown up during the Depression and he knew what hunger was. There had been times he and his nine siblings had had little to eat, but never nothing. It had been nearly a day since Betty Mae, Gladys or any of the Maule children had eaten anything and Paul doubted that they were the only children in that situation.

Paul asked Osborn to walk his sister home in case she fell again. As he looked out over the nearly two dozen faces staring back at him, Paul realized that it wasn't just the Maules who were starving. It was all of them. All of the students in his charge were starving to death. Not even all of his students were present and he doubted that Betty

Mae would be here tomorrow. Were the absent children like Betty Mae? Were they too weak to sit at their desks like Gladys?

How can someone watch another person die and not be affected?

Paul knew the answer. *He doesn't want to see it.*

But that didn't mean that once he did see it, he had to sit back and let it happen.

Gladys Maule

9

A Town in Need

Paul spent the rest of the school day carefully watching the students and not just his students in the Big Room. He watched all of the boys and girls in the Little Room, too. He wasn't going to have a repeat of Betty Mae and Gladys. Not today. Not ever.

He found himself counting the number of bare feet he saw shifting uncomfortably on the cold, wood floor of the school. It wasn't only food the children needed but clothes, too. He counted holes in their clothing. The boys' clothes were in the worst shape. Their pants had holes in the knees and torn pockets. Their shirts were ripped from roughhousing or snagging on branches in the woods. Some had been patched and repaired and even the patches had worn out. The girls' dresses fared better. They didn't have holes, but their dresses were drafty. Paul knew from things Jerry had said that many of the boys no longer wore underwear. With scarce resources, underwear became something reserved for the girls and their drafty dresses.

Though there was no snow on the ground yet, it would come. They were in the mountains after all. And there had been frost on the ground in the mornings lately. These children had walked through it to come to school without shoes and coats.

They shivered, but they didn't complain. Paul took small solace that as long as they felt the cold, they weren't frostbitten. It was a small comfort. It was like saying, "I'm going to slice you to death with a knife, but it's been sanitized so you won't get an infection."

A more-immediate comfort was that no one else fainted, though some of them looked as if they might. How many of them hadn't eaten anything for lunch while he had sat at his desk barely tasting his own sandwich and slice of pie as he graded papers? Paul was sure that his face burned with shame.

At one point, he left his students working on an assignment and walked across the hall to the Little Room where Elva Mae Dean taught the younger students. Though Paul was a teaching principal at the school, Elva had more experience in the classroom. She was a middle-aged woman who was slightly overweight. She generally wore her hair short and had a friendly smile, especially for her students.

Elva took educating her students as seriously as Paul did and he liked that. Once, she had failed her entire first grade class except for one student. Since there had been less than ten first graders and they would have stayed in the same room for second grade anyway, it wasn't as dramatic a gesture as it sounded. Still, she made her point that the students needed to pay attention and do their assignments.

Elva Mae Dean

She had taught at Shallmar School from October 1943 to June 1947 only to return in December 1947. So she knew this town, or at least its students, better than Paul. She would know if something was wrong.

When he asked about her students, though, she had no real concern. She said they had been growing more lethargic, but that also allowed them to concentrate more on their lessons. No one had fallen over in her classroom. At least not yet.

Paul also noticed that a higher than normal number of students were absent from her class. All told, he figured that he might be missing a quarter of his students if not more.

He wondered how the high school students were faring. They had a two-mile walk along Shallmar Road into Kitzmiller to get to school. If they weren't eating much, that was a walk that would exhaust them before their school day even started.

Now that he knew something was wrong, Paul couldn't help but see the signs of it all around him. The lethargy that Elva mentioned, the increased absenteeism, the growing thinness in his students. And now that he had seen it, he couldn't help but think about what could be done to fix things.

The obvious solution was that the families in Shallmar needed food. Now, the Bible talked about teaching a man to fish as opposed to simply giving him fish to eat. For coal miners, that meant they needed to be able to work so they could be paid. Then they could take care of themselves and their families.

The mining company wasn't going to be any help since it had closed the mine, but the miners' union probably wouldn't be much help either. The Shallmar miners hadn't even participated in the recently ended national strike.

So what did that leave as an option?

After school dismissed for the day, Paul walked across the baseball field and around the side of the company store. Rather than cross Shallmar's main street and walk to his house, he walked past it. The unpaved road paralleled the North Branch Potomac River, which ran only a couple dozen yards to the south of the street.

The houses in Shallmar had once been considered among the nicest homes for miners in the region. Now a lot of them needed a fresh coat of paint and more than a few were boarded up because they were no longer livable or needed. Shallmar had large homes that were two stories with basements and three rooms on each floor. However, only four of them had bathrooms and one of them was Paul's house. He had lived in a house without an indoor bathroom for about a month after he and his family first moved to Shallmar. He had been more than happy to pay the additional rent for indoor plumbing when his current house had become available.

The smaller houses in town were four-room bungalows. They had a lower rent, but they had no indoor plumbing.

The Wolf Den Coal Corporation had tentatively agreed to put bathrooms in each of the houses, but that had been before the drastic downturn in business. As a result, the company had decided to offer the homes for sale, but the miners couldn't afford to buy them. By 1949, most of the miners couldn't even afford to pay the monthly rent of $7 to $15. Some of them were more than a year behind in their payments, though Paul didn't know it at the time.

Because of the size of the Maule family, Paul had expected them to live in one of the large houses on the main road. He was surprised to find that they rented a smaller bungalow on one of the cul-de-sacs.

Catherine Maule invited Paul into the crowded house, which consisted of a living room, kitchen and two bedrooms all on one floor. To say that there was 800 square feet of living space inside the bungalow would have been generous.

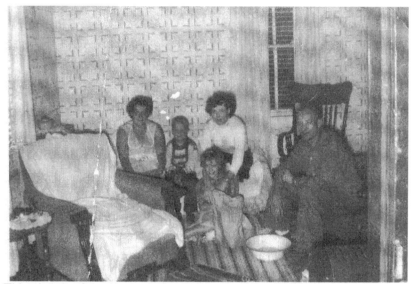

This picture of the Hartman family in one of the Shallmar company houses show how small the rooms were and these houses were considered among the best company houses for miners anywhere. Photo courtesy of Robert Hartman.

The bathroom was a tub that hung on the back wall of the bungalow and an outhouse sat further back to minimize the smell. The only sink was in the kitchen and that had to be filled with water from a nearby community spigot. In the winter, the water might be heated on the stove before being used to fill the tub, but Shallmar's families grew used to cold baths in the spring and summer. That wasn't unusual.

The house also didn't have electricity, but then, neither did any of the houses in Shallmar since the mine shut down.

Walter Maule was a forty-five-year-old pony driver for Wolf Den Coal and had been for thirty-five years. He had only worked twelve days for the mining company this year and only five days since the mine had closed. Although he wasn't a miner, Walter was out of work just like everyone else who worked in the mines. If anything, it was probably harder for him to find work. There were fewer positions for pony drivers than miners and the more the mines modernized, the fewer jobs there would be.

"I came by to check on Betty Mae and Gladys," Paul said to Catherine once she had shown him inside.

She had been reluctant to do that, but otherwise, she would have had to talk to him on the stoop. Once Paul had started asking about how the family was doing, he had been shown inside. Paul doubted it was because of courtesy, but rather that the Maules didn't want the neighbors to hear about their business.

The entire Maule family was inside the cramped bungalow not that they had anyplace else to go. No one was working. School was out for the day and the temperature outside was in the teens.

The Maules had six children—Betty Mae, Gladys, Osborn, Joseph, James and Minnie. Betty Mae, Gladys and Osborn all attended Shallmar School in Paul's class so he knew them fairly well. Joseph attended high School in Kitzmiller, a two-mile walk from Shallmar. James had graduated and gone off on his own to find work.

The oldest Maule child, Minnie, was twenty-one years old. She had married and was now Minnie Hedrick. She was also already a widow after her husband had been killed in a mine accident. Minnie and her young son, Walter, who was named after his grandfather, had moved in with her parents because she had no place else to go and no savings. With her husband dead, she couldn't remain in a company-owned house.

The house was well kept, but then, there wasn't a lot to keep. Although the pot-belly stove was burning in the living room, it was still chilly near the doorway where Paul stood. At first, he wondered if the fire had burned down, but then he noticed the newspaper pages taped across the windows and on the walls. It was a form of insulation. Not a very effective one, but when you had nothing, every little bit helped. Even Paul's house—one of the nicer ones in town—had no insulation. The walls were plaster and lathe on the inside and clapboard on the outside. None of the houses in Shallmar was ever completely warm in the winter.

Paul saw Walter Maule seated in a chair near the stove. Paul walked over and shook his hand, but the middle-aged man didn't rise out of his chair. It was a somewhat rude treatment for a guest, but Paul had his doubts that Walter Maule could have stood up. His hand felt bony and his cheekbones pressed against his skin. Walter's grip was weak and he looked tired, even a bit confused.

Paul had trouble matching the man in front of him with a coal miner. When mines were operating, miners ate huge breakfasts of potatoes, gravy, pork side and more. By some estimates a miner's breakfast could be five times the size of breakfast nowadays. They burned up those calories doing hard, physical labor in the mines. Miners tended to be lean and strong, but Walter had a look that reminded Paul of the pictures of American prisoners of war in the Pacific. He had that same haunted look to his face.

Paul took a seat and asked what was wrong, explaining to Walter and Catherine that he knew Betty Mae hadn't eaten that morning. Catherine was the one who responded. Walter didn't look like he was in any shape to do so. Her explanation was simple: The mine had closed.

When Paul pointed out to Catherine that it had been eight months since the mine had closed, she explained that when the Wolf Den Mine shut down, the miners had been able to collect unemployment benefits, but those had only lasted until July or August if the miner had put off filing for as long as he could.

Albert Males, the chairman of the United Mine Workers local relief committee, had started giving Shallmar families small relief checks from the union's treasury. He only had $1,000 to work with and that was only because he cashed in all of the union's savings bonds. Split among four dozen families that money wouldn't last too long, especially when it wasn't being replenished with dues. Each family got $2 to $7 a week based on the size of their family.

Albert hadn't told the national union of the situation to tap into its welfare fund because as he explained it, the national fund was for hospitalization and pension payments not unemployment payments because a mine had closed. If it had been, then the international fund probably would have gone bankrupt years ago.

Paul realized that he knew or had heard most of that information, but it hadn't all come together as to what it meant until he saw Betty Mae faint. Catherine told him that for most of the past week, her family had been eating only apples that the children had found at Wilson's Farm up on the mountain. The apples had been frozen and chances were that they had made the Gladys and Betty Mae sick as much as the lack of food had made them faint.

Not that that would be a problem again. The apples had run out the day before, which is why Betty Mae, Gladys and all of the Maules, hadn't eaten that morning. It wasn't the first time they had gone without breakfast lately either.

"My children have forgotten what milk tastes like," Catherine Maule said.

Miners like these were the life blood of Shallmar. When the mine closed, most of the jobs in Shallmar disappeared. Courtesy of Western Maryland's Historical Library.

A lack of food was also why Walter hadn't been able to stand up to shake Paul's hand. He was missing more meals than anyone else so there would be more food for his family.

Catherine assured Paul that they had managed to find enough cabbage and potatoes to get them through the next day. She was proud of this because, "We sometimes don't even have potatoes and cabbage," she told Paul.

When Paul told her that they needed more than vegetables, Catherine told him, "We haven't had any meat or milk for four months."

Paul asked what the baby ate if the family wasn't getting milk. Minnie told him, "We get a can of condensed milk every other day."

The news shocked Paul, though perhaps it shouldn't. The diets of families living in coal mining towns could easily find themselves lacking some basic food items if the miner was out of work or working very little. Flour, corn meal, beans, potatoes and bacon were staples in a miner's diet. Forget drinking fresh milk. Like the Maules, even families with small children gave milk up in favor of canned milk. Fresh vegetables were absent in the winter and spring. Unsalted meat was a treat if it was available at all.

What Paul had just learned led him to believe things were worse than he had imagined. Shallmar was a small town with less than sixty families and maybe 200 residents altogether. How could neighbors and friends let this happen?

When Paul asked about it, Catherine explained that when her family had gotten down to one slice of bread for all of them, their neighbors had asked her how they were doing. Knowing that the neighbors weren't much better off than the Maules, Catherine had told them that they had food for four or five days.

How close to starvation a family was was not something people generally talked about so to talk about it wouldn't be checking on how a neighbor was doing so much as comparing degrees of misery. Shallmar was a coal town and a company one at that. Nearly everyone in town depended on coal coming out of the ground.

The store owner, Baxter Kimble, depended on the mine. If miners weren't earning any money, then they weren't spending anything in the company store. Even Howard Marshall needed the miners working. If they weren't, then he wasn't making any money from selling the coal.

Even Paul, who was paid by the Garrett County Board of Education, could see his job disappear from a lack of work in the mines. If the miners weren't working, they might move from the area, which would decrease the number of students in town and the need for a school.

It was all about people getting paid to mine coal and in Shallmar, they weren't.

10

Calling For Help

When Paul left the Maules and walked home, he felt like he had the weight of the world—or at least, one small coal mining town—on his shoulders. He was having trouble getting his arms around the situation because it seemed so impossible.

How could a town be starving? How could the people here have let things get so bad? Even as the thought formed, he knew the answer.

A frog got cooked.

If you want to cook a frog, you didn't throw it in a pot of boiling water. It would jump out as soon as it hit the water. No, to cook a frog, you put it in a nice, cool pot of water and gradually turned up the heat. By the time the frog realized there was a problem, it was too late.

So it was here. Shallmar was the pot and the diggers and their families were the frogs. The heat was the buildup of lots of things that by themselves would not be too bad, but together they were devastating.

The mine had closed. It happened time and again, but this time of year, it had followed another bad year so the miners had already been living paycheck to paycheck. No work meant no income. It also meant that any future unemployment benefits would be reduced. Because the Shallmar miners had had so little work in 1948, their unemployment checks were less than half of what they could have earned working a three-day week.

When Catherine Maule had told Paul this, he replied that there were other jobs. "Not in Shallmar" had been the reply and since nearly everyone in town was too poor to afford a car most of them couldn't drive somewhere else to work. The diggers were locked into the short radius of where they could walk to, which pretty much meant Shallmar, Dodson and Kitzmiller and Kitzmiller was the only one of the three that had any businesses beyond a company store. Sure, a person might be able to hitch a ride to somewhere further away, but you couldn't depend on hitchhiking to get to a new job.

They couldn't move to a new town. No one had any money to afford the move or rent a new place to live. Three families had moved out of town after the mine had closed. It appeared that they were frogs who had felt the heat and jumped.

Everyone else was trapped here.

The whole situation left Paul feeling guilty. He had a car. He had a job. He only had one child to feed unlike the Maules. His situation was nearly the exact opposite of theirs.

The miners didn't complain, but then, there wasn't much that they could do. They couldn't help themselves, but he could. He had to before something worse happened.

Maybe the attitude of the miners was getting to him. You don't whine. You endure. You don't give up. You press on. Miners were tough people. Their lives demanded it.

But Paul wasn't a miner. He worked with the children of Shallmar and Dodson. It was often a thankless job, which he could endure. But to have to look into the faces of those children each morning and wonder which one would be the next to collapse was more than he could handle. Paul doubted the toughest miner in Shallmar could have managed that.

Shallmar was barely over thirty years old and it was dying. The sad thing was that many of the old timers in town said it had been going on for a while, since the tipple had burned down in 1930. Paul hadn't believed them, but now he figured he owed all of them an apology.

He walked along the shoulder of the road more out of habit than anything else. Shallmar had no sidewalks, which didn't matter much since there was also very little traffic to endanger anyone walking on the road.

The Andrick house was one of the nicer ones in Shallmar. Molly made a point of washing the coal dust off whenever it got too dirty. It used to be a weekly chore, but come to think of it, she hadn't had to do it all summer and fall.

He walked in the front door and put his coat on the hook beside the door. The front door opened into his den. It wasn't a large room, which is why it had relegated as the den, but it was large enough to hold his desk and chair, radio set and bookcase.

Turning to his right, he walked into the living room and sat down in his easy chair. Though the living room was larger than his den, much of the space was taken up by his wife's piano. Molly loved to play and sing. Paul would just listen to her. He couldn't carry a tune in a bucket. The other large piece of furniture in the room was the sofa, which like his easy chair was covered in a plastic slipcover to help keep it clean of coal dust and child dirt.

He could see his wife through the doorway in the kitchen putting the finishing touches on their dinner. He called out to tell her that he was home and then he let himself lapse into his thoughts.

After a few moments, Molly walked out and sat down on the sofa. When she asked about his day, he told her. He told her everything and he wasn't sure how he kept from crying.

When he finished, she asked, "What are you going to do?"

That was the question that had been bothering him since he had seen Betty Mae faint and Gladys nearly do so. He now knew the answer. He just didn't know if his solution would help. He was one man with little power that would help. He had to let more people know and people with power so that something could be done.

Paul told his wife this. Then he went into the den and sat down at his desk to write letters to Garrett County Board of Education, the county's representatives in the Maryland Legislature, the county's congressman, two U.S. senators and county commissioners. It was their job—their responsibility—to help the county's residents. His letter to the school board noted that absenteeism in his school was on the rise not only because children didn't have anything to eat, but they didn't have proper winter clothing.

Paul had noticed that four of the Crouse children had been absent from school lately because their shoes had been worn out and they had nothing to wear on their feet.

He also wrote to the state office for unemployment security in Oakland that was in charge of unemployment benefits for workers in Garrett County. Paul explained in the letter the situation in town and asked for whatever help could be found, such as government surplus food or contracts for the mining company.

Paul was still writing when Molly called him to dinner. He ate without enjoyment because he couldn't help but wonder who in town wasn't eating. He would look at Jerry and wonder how he would feel if Jerry had nothing to eat. He would probably feel like Walter Maule even to the point of giving up his own food.

After the meal, he finished the letters. Then he wrote one last letter to *The Republican* newspaper in Oakland. He had let the people with power know the problem. This last letter would let a lot of people know. Individually, they might not be able to do anything, but together, they could press for help.

Having done all that he could for the evening, he went to bed. That's not to say he slept. Paul tossed and turned. His head was filled with all kinds of thoughts; some wild, others focused on problems. If Shallmar had been a wounded person, the first thing he would do was stop the bleeding. Then he would get care for the person.

So what was the bleeding in Shallmar? What was the immediate need that needed to be treated?

None of us have eaten.

Food. The people of Shallmar needed to eat. With food, they would have the energy to work if it presented itself and at least they wouldn't starve.

When Molly made his lunch the next morning, Paul stood beside her and made an extra two sandwiches and added in some extra apples. She didn't make a joke about his appetite. She knew what he would do with the extra food.

During the lunch break at school that day, Paul watched the children in the Big Room. He noted that there were five children who were without lunches or with very little to eat. Paul sliced up the apples and sandwiches and set them on a book. Then he walked down the aisles of desks handing out the sandwiches and apples, including his own lunch, to the children who needed a meal.

He tried to be nonchalant about it. He didn't want to embarrass the children. Some of them had as much pride as their parents, but they were also hungry. They would smile at him and wait until he passed so that the eyes in the classroom followed him and didn't linger on them. Some of the students would nibble at the meager meal hoping to make it last. For them, it was a fight between hunger and patience. Others would simply give in to their hunger and wolf down their food.

Paul gave away the food until he had nothing left even for himself. Then he sat down at his desk and went back to grading papers.

He could feel the change in the classroom without even looking up. Murmuring increased. He heard more scraping of chairs and desks on the wood floor. He even heard some giggling. It sounded like a normal classroom.

Class went well for the rest of the day. The students were more attentive and more willing to volunteer to answer questions. Paul was so pleased at the change in the mood of the class that he didn't even notice his own hunger until school ended. It didn't matter. It was a small price to pay if the children were no longer hungry.

After the students had gone, he left the school and headed across the baseball field to the company store. It was a large one-story building made of gray block stone. It was small by the standards of many company store but since it was the only store in Shallmar, it was also the largest. Baxter Kimble ran the store, but he leased the building from the coal company.

The children of Shallmar School. Photo courtesy of Jerry Andrick.

The road from Kitzmiller was paved up to the store, but from there into Shallmar, the road was unpaved as if to say that civilization stopped at the store.

The store itself had no identifying sign on the outside, but it didn't need one. It was the only business in town, other than the coal company, and the only building that looked like it did.

The building had two entrances. On the right, was a door that led into offices for the Wolf Den Coal Company that took up about one-third of the building. This was the payroll office for the miners. The miners would line up on pay day and pick up their earnings at a window on the side of the building. The company paid miners with cash or scrip so besides a safe in the office, you could also find a shotgun and billy club in case there was trouble.

After they miners got their pay, they could walk around the corner and through the door on the left into the company store.

Inside the Shallmar Company Store. Photo courtesy of the Garrett County Historical Society.

Inside the store, which was heated by a large pot-belly stove in the middle of the room, was a post office and a general store that sold all the bare necessities for living in Shallmar if you could afford to pay for them. The prices in a company store were ten to fifteen percent higher than you would pay at the stores in Keyser or Oakland.

People who could go to Keyser or Oakland to shop in their markets and multi-story department stores with bright lighting, helpful salespeople and enticing displays would find the Shallmar Company Store a disappointment. But for miners who infrequently got out of town to shop, it sufficed. Moreover, miners were expected to shop as much as they could at the store, which was owned by the mining company that supported them. When a miner didn't, somehow word always got back to Howard Marshall or Jesse Walker. Some miners were even told of how Howard Marshall had told them that they would be expected to shop at the store as a condition of their employment with the mining company.

Shelves lined most of three walls except where the post office boxes were. The same was true for the counter and display cases that ran around the room. Most of the center of the store was empty except for the pickle barrel filled with large dill pickles.

Miners would come into the store on cold days and sit around the stove and talk. On warm days, they would sit on the benches by the ball field, chew tobacco and talk. When kids from town would wander over, the miners would spit tobacco juice at their feet to send them on the way.

Paul handed the letters he had written the previous night to Baxter Kimble, the store manager and postmaster. Baxter put them in a small bag of outgoing mail.

Paul turned away and watched the miners around the stove. Usually, they would be talking and joking, but this group was fairly quiet like Paul's students had been this morning.

How much longer would it be before they started falling over? Walter Maule had looked like it had already happened to him.

Paul walked home, wondering what more he could do. He was feeling frustrated and powerless. He sat down on the sofa in the living room.

"What if no one can help us?" he asked, letting his frustration flow freely.

"There has to be someone," Martha told him as she sat down next to him.

"Then why hasn't he helped already? These families need help now."

"Then we'll help."

"We can't support a town."

"No, but we can help some of them like you did today."

"It's not enough. It won't make a difference."

"It will to the ones we help."

11

Finding Her Place

Molly Andrick was never quite sure where she fit into among Shallmar's society if you could call the group of hard-working women in town a society. It was certainly nothing like you read about in the *Cumberland Evening Times* with women going to luncheons to hear a speaker while they nibbled on finger sandwiches or dressing in lovely gowns to attend a charitable ball.

Now the women of Shallmar were friendly enough, and they tried to include Molly and make her feel welcome, but more often than not, she felt like she wasn't contributing much. She wasn't a coal miner's wife even though she lived in a mining town, and so she saw things differently than these women did.

She did like the fact that the women acted like partners with their husbands. Each one had a job to keep the family functioning and they did their jobs as best they could.

The women in Shallmar had no household help to wash their laundry and prepare their meals and they could have used the help more than Cumberland's society women. Trying to keep a house, even one as small as the bungalows in Shallmar, free of coal dust was a full-time job. Add all the other chores that needed to be done to their day and there was no way to keep up.

Not that the wives complained, though. At least they could see the sunlight and death wasn't looming over their heads every second they were at work. They didn't have to worry about walking down a hallway and falling over dead because there was no air or developing the hacking cough that came from inhaling coal dust for hours each day.

Besides, these women had been born in coal towns and raised in them for the most part. Their daddies had been diggers and so were their brothers. Their mothers had even gone into the mines to work during World War I. This was how life was. They didn't know any other way.

But Molly did.

Yes, she had grown up during the Great Depression and she had known hunger. She was the youngest of eight children who all had to be supported on a mechanic's pay. Although her father, Granville McDonald had owned his garage in Mayville, his customers had often paid him with moonshine because money was so scarce. That meant their home had gone through periods with no electricity, leaky roofs and no indoor plumbing.

Martha "Molly" Andrick

But Molly had lived in towns large enough to be called cities and she had graduated college. Yet, here she was in Shallmar trying to make friends with these women who weren't relatives or childhood friends and with whom she did not share a common background.

It frustrated her. She didn't feel like a part of this group of women and she couldn't use her college degree. Even if there had been a teaching job available, she wouldn't have been able to take it. Not if the job was in Maryland. She supposed that she could drive over the Kitzmiller bridge into West Virginia, and maybe find a job a one of the schools in Grant or Mineral counties. That would just mean a lot of driving in bad weather on poor roads that would just add to her time away from J. Paul and Jerry. She needed to find a job closer to home, though that was what everyone was looking for and something Shallmar, Dodson and Kitzmiller just did not have.

So until that time came, Molly needed to find her place in Shallmar. She tried to find common ground with the women. Sometimes she would sing or play the piano with them, but finding the time to do it was a challenge. The women's daily chores pretty much filled their days.

Molly got up in the mornings with J. Paul to make him and Jerry breakfast before they headed off to school.

During the day, she would work in her garden if there was work to be done. If the vegetables had been harvested recently, she might spend most of the day canning them in jars she could buy at the company store. One of Jerry's chores was to pick berries when they were in season, which she would also can.

She baked fresh bread once a week, at times, trading recipes for different types with some of the women. Some of the women baked more often, but they had larger families who went through bread quick. If Molly baked more than two loaves a week, some of it was bound to go stale. When it did, she mixed in some butter and tomatoes to make a type of bread pudding.

She did her wash on Mondays along with the other women. A few of the women had gas-powered washing machines, but most of them filled large tubs with water from one of the community spigots and scrubbed their family's clothes on washboards while trying to avoid skinning their knuckles on the board as they worked to get the coal dust out the clothes. Molly had a washing machine to help her wash clothes, but when they had first moved into town, they had lived in a house without water for about a month until their current house was available. During that time, she had scrubbed clothes in a tub like the other women in town and she was doing so again since there was no power.

Wash day also meant that all of the throw rugs in the house were washed in the tub as well. The rugs were washed until they were free of coal dust and run through a hand-cranked wringer and left out to dry.

Once a week or so, Molly walked to the company store with an order for some of the basic items. The store carried most of its food items in bulk, which had to be measured out by Baxter Kimble or one of his clerks. The store was a good enough place to buy things like sugar, lard and flour. Canned goods and non-food items were also safe to buy, though more expensive than if she bought them in Oakland or Keyser. Kids would come in to buy moon pies and orange Crush or they might buy penny candy from the containers on the counter or candy barrel.

The store had a walk-in refrigerator at the rear. When a customer ordered meat, Baxter would walk back there grab the meat he needed and grind it himself and put it in the meat display case. Baxter also had a yellow cat that he allowed to roam the store, walking on the counter where food might be laid.

Thankfully, she wasn't forced to buy any perishables at the store. The vegetables might be fresh if you were there when they were brought into the store; otherwise, they usually looked fairly old. As for meat, it was something you didn't want to buy there even on the rare instances when it was available. The store had no way to refrigerate anything when the mine shut down so it tended to go bad quickly.

Because the families in town were so limited by what they could eat to either what they could grow in their gardens, hunt in the mountains or buy in the company store, Molly quickly learned that a miner's diet differed from what she tried to cook for Jerry and J. Paul. Miners ate a lot of beans, potatoes and fresh game. Molly tended to add a lot more vegetables to her meals.

For all her feelings of being an outsider, Molly was probably fitting in better than her husband. J. Paul had been shocked when Betty Mae Maule fainted at school. That surprised Molly because she would have thought that he would have seen what was going on around him being at the school every day.

Molly had seen the destitution creeping into the town. Things had already been pretty bad when they had moved to Shallmar, and it had only grown worse with the past year. She had seen it with small changes after the mine closure. The only money they had gotten from the coal company since then was about $80 in June. It was called vacation pay, though it served more as severance pay for the miners.

The women washed fewer clothes and didn't clean as much because the mine wasn't putting any coal dust into the air. The conversations among the women began to avoid things like food and money.

As summer progressed, Molly had seen fewer and fewer of the wives in the company store. She also saw them enlarging their gardens and gathering more fruits and other things to eat, some of which Martha wasn't sure she could bring herself to eat like rattlesnakes, rock lettuce, milkweed and poke.

She also saw that the clothing that children wore became more worn out with larger holes in the knees and rips that weren't mended.

People in town also looked more tired and worn out, though they weren't working as hard. They were suffering and they didn't want to talk about it, but in not talking about it, it seemed to ensure that everyone knew.

J. Paul wasn't like that. He spoke his mind. There had been an incident once in Red House School when J. Paul had been teaching there that left Martha surprised J. Paul still had a job. One of the county education directors had visited the school and started telling J. Paul all of the things he needed to do to run his school better. Now, mind you, this director decided on all of this after only a couple hours at the school, whereas J. Paul had been in charge of the school for months.

So, J. Paul asked the man, "What type of college degree to you have?"

"I have a bachelor's degree in education," the man said proudly.

"Well, I have a master's degree in school administration so I think I know what I'm doing better than you do."

Even more important than speaking his mind, J. Paul understood people. Martha wasn't sure if he was born that way or whether he learned to read people on the football field. Not that it mattered. Now that he was paying attention and saw a problem in Shallmar, he would do something.

He knew when to accept a burden as doing his share and when to call for help because the problem was too big to handle on your own.

J. Paul was going to call for help.

Shallmar School. Photo courtesy of Jerry Andrick.

12

On the Brink

Not one to wait when there was work to be done, Paul continued moving forward even if it felt at times that it was like trying to cut a lawn one blade of grass at a time. At school each morning, he would take a tally to find out which students had something to eat either for breakfast or lunch. Molly would come by later in the morning to get the tally from her husband. She would walk back home and make lunches for the students who didn't have anything to eat. The lunches would include a sandwich, a cup of milk and perhaps a piece of fruit, if the Andricks had it.

One afternoon in early December 1949 a stranger came to town. Paul didn't see him at first because he was teaching at the school. However, the man stopped at the company store to ask directions to Paul's house from the miners sitting around the store's pot-belly stove. The men were polite as they smoked hand-rolled cigarettes, chewed tobacco and tried to keep warm. They sent the man off in the right direction. The Andricks' house wasn't hard to find since it was only a couple of houses further down the street.

As soon as the man left the store, the miners started speculating on who he was. With no work in town and any available newspaper being used for insulation rather than reading, the men in the store had worked their way through most topics of conversation by nine-fifteen in the morning. The stranger's appearance gave them something to talk about for another hour.

Then they left to tell their bit of real news to anyone who they could find.

By the time Paul left school and headed to the store to check his mailbox at the company store, he was one of only a few adults in town who didn't know that there was a stranger at his house. No one at the store bothered to tell him either. Who were they to ruin the surprise Paul had waiting for them? Things might get as interesting as when Howard Marshall was cheating on his first wife. Paul's reaction might make another piece of news that they could spend some time musing over.

So Paul walked into his house and was surprised to see Molly sitting in the living room talking to a stranger. The man introduced himself as a reporter with the *Oakland Republican* newspaper who had been asked to follow up on Paul's letter to the editor. The two men spoke for a bit and then Paul decided that it would be better to show the reporter the problem.

They went outside and walked down the road towards the turn to the mine. The reporter could see for himself the situation and condition of the once-immaculate homes of Shallmar; the bright whitewash faded, paint peeling, rose bushes grown wild and untrimmed. Some houses were even boarded up, having fallen into disrepair because it had been so long since anyone had lived in them. Paul had seen Oakland and sometimes shopped there if his family didn't head to Keyser. Though Keyser was further away, Martha had family who lived in that city so the trip served two purposes. They'd make a day trip of it; visiting family, shopping and sharing a meal.

While Paul and the reporter walked along the dirt road, they talked. Paul didn't use anyone's names. If the families wanted to talk with the reporter, well, that was their own business. Otherwise, Paul would respect their privacy. He spoke about the problem in generalities. He didn't say that Betty Mae had collapsed. He said his students weren't eating regularly and that one had fainted and others had come close to doing so.

"Without a great amount of help from the outside, these people cannot hope to survive the winter," Paul told the reporter.

Anyway, if this reporter lived in Oakland, then Shallmar probably looked like a throwback to the turn of the century. It had no street lights and no paved road. The houses had no power. Few of them had telephones or indoor plumbing.

The reporter asked his questions, wrote down his observations and the answers to Paul's questions and then headed back up the side of the mountain in his car. Paul could do nothing now but wait and see what fruit his visit bore.

The men in town looked forward to deer season, especially after wild turkey season in November had been disappointing. The season ran from December fifth through the ninth. A huge buck could feed a family for weeks, not only with fresh meat but also dried venison. The pelt could also make a blanket or rug to help a family stay warm in the winter or sold for a few extra dollars that could feed a family for a couple days.

Out of work miners turned to hunting to keep their families fed at Shallmar. Photo courtesy of Bob Hartman.

The men even had plenty of time to hunt. They marched out into the woods early in the morning and looked for a solution to their hunger. At the end of the day, nearly every one of them returned home empty handed. Many of Western Maryland's deer, particularly in Allegany County, were dying from a "toxic poison" in 1949 and that severely reduced the number of targets for hunters. Hunters from Shallmar only managed to bag four deer or only four that were legally recorded for the 1949 season.

"I never cared much for venison, but it was the first fresh meat in this house for three months," one woman told a reporter.

George Brady bagged a 190-pound buck during the season. It came in handy since the Bradys had cooked the 100 chickens they had owned as their food stores had dwindled. They had eaten many of them, but they had also used them to feed friends and family.

Since the nearest recording station was in Swanton, Maryland, and nearly all Shallmar residents were without a car, some deer may have been shot that were never recorded. Desperation drove people to do what they ordinarily might not do. Take the Hanlins, for instance. Not only was Robert Hanlin out of work, but what money he did receive tended to go towards buying a pint of moonshine. So if the Hanlins expected to eat, they needed to find their own food. They hunted anything, whether it was in season or not. If it was edible, it was in season. Other men did it, as well. They would eat anything – rabbit, squirrel and groundhog. One family even had a recipe for pickled groundhog.

Bill Crouse's father, John, was one of the lucky hunters who bagged a deer in season, but it wasn't enough to keep his family fed for too long. He had been a coal miner in Shallmar since the town had been built through two world wars, the Great Depression and who knows how many strikes, and this was the first time that he hadn't been able to support his family.

"It was the first time we had fresh meat in eight months," John's wife, Dolly, said after her husband brought home his deer.

No more was she running into the company store to spend every cent of her husband's pay. He had no pay to spend, not in cash or scrip, not even enough for a penny piece of candy.

The Crouses were long-time residents of Shallmar and all six of their children had been born in town with the assistance of one of the midwives in the area. Mildred Amy Sharpless and Amy Pugh helped more coal miners' children into the world than did the doctors in the area. Not only were midwives cheaper to use than a doctor, but midwives usually lived closer to the mother who was in labor. Doctors tended to live out of the area,—definitely out of town—which meant that someone would have to go find a phone on which to make a call and then the doctor usually arrived after the baby was born. Even when a local doctor was called, he tended to be a late arrival on the scene.

As I said earlier, diggers are a tough lot used to doing without or making due on their own. This quality extended to their families as well. You had to be pretty sick to get out of doing your part for the family and even then, you didn't go to the doctor.

"We doctor ourselves," one unemployed digger's wife said. "Every time you get a doctor, it takes that much off what you have to eat."

Families tended to rely on home remedies in order to save money. A couple drops of kerosene in a spoonful of sugar supposedly help cure a cough or sore throat and sheep manure tea was used to bring down a fever.

The three oldest Crouse children—John Jr., Theodore and Robert—attended Kitzmiller High School. Christiana was a student in the

Big Room in Paul's class and Bill was a classmate of Jerry's in the Little Room. The youngest Crouse child, Charlotte, wouldn't start school until the next year.

Bobby Crouse

From his asking around to find out how bad Shallmar's situation was, Paul knew that the last of the unemployment checks for most everyone had run out in August. Dolly, who could be a spendthrift when the money was available, also knew how to pinch pennies. She had made her family's money stretch out longer than most. They had one meal a day on many days and that meal was potatoes and baked beans.

"Today for a change we had cabbage for supper," she told one newspaper reporter.

The supplemental union help, three dollars a week, had ended shortly after the unemployment checks stopped, and Dolly had stretched that out as well. She didn't buy pre-made foods. She repaired anything that ripped or broke.

Being so miserly was having its effect on her family. Her three oldest children had stopped attending school because they had no shoes to wear in order to make the two-mile walk to Kitzmiller on the freezing cold road. Paul had seen students at his school going barefoot or wearing shoes that had cardboard inserts for soles because the original soles had holes in them.

Yes, an education was important for her children's future, but she couldn't afford new shoes or even soles for the old shoes. She also couldn't afford a doctor's visit if one of her kids got sick.

It was times like these when the partnership between a coal miner and his wife became obvious. Each one had their duties. Usually the miner worked his long days while his wife managed the house and kids. During a strike or mine shut downs, managing the house became even more important when money was tight.

Christiana Crouse

To make matters worse for the Crouses, their oldest son was in the hospital. Seventeen-year-old John Jr. was a diabetic so his family was careful about what he ate. However, during blackberry season, hunger had gotten the better of him. He and some of his friends had climbed up Backbone Mountain and found a blackberry patch and then proceeded to gorge themselves. The result was that John Jr. was now eating better than his family as he recovered from an insulin shock.

Every family in Shallmar was suffering to some degree if for no other reason than no work meant no pay. Those families who had managed to save some of their pay from the good times fared somewhat better, but after two lean work years, their savings accounts were now as empty as their bellies. Coal mining had never been a job that paid a lot and miners rarely worked full weeks since the war ended.

Dixie Crosco, a town resident, told a reporter, "We ran out of food and money last week. What are we going to do? Dolores here is six years old and she's a good girl. But for several weeks I couldn't send her to school because she didn't have any shoes. She didn't have enough to eat, either."

Dolores was Dixie's daughter from a previous marriage.

Norma Jean Crouse

Paul knew about that situation because Dixie had told him that she wasn't going to send her children to school barefooted in the winter. Not only had she started keeping Dolores out of school but also her brother, Thomas Richard who was seven years old.

"I never saw things as bad as they are here now—not even during the Depression," Tony Crosco, Dixie's husband, told a reporter.

He had tried to collect wild ginseng to support his family because, dried, ginseng was highly valued for its medicinal properties and sold for $13 a pound. The problem was that he could find very little of it. When he applied for relief, he was told that he should get his sons to support them as was often done among families. Four of the five of them were coal miners, though.

"I can't ask them to help," Tony said. "They are married and have their own families to support and they aren't any better off than I am."

Paul would have helped everyone if he could have afforded it. Their stories tore at his heart. His feelings ranged from sympathy for the family to anger at the lack of help from the people who should have helped to frustration with himself over not being able to help.

Besides feeding students at school, Paul dropped some food off with the Maules. The consensus was that they were the family who was the worst off in town and having learned even more about their problems, Paul didn't doubt it. Their home had only one stove in it to heat the entire bungalow. Although they lived in a coal mining town, to get coal for their stove, Walter Maule had to scavenge coal from an abandoned mine several miles away. That meant that he and the children had to walk a couple times a week to the old mine to fill buckets with dirty coal and then walk back to Shallmar carrying those full buckets. It was exhausting work, even more so, when they had so little to eat.

The living room of the Maule home had a few pieces of battered furniture in it and laundry hung on lines strung across the room to dry. The house was perpetually dark because newspaper also covered the windows. The house's light came from a single kerosene lamp that was used primarily at night. This meant that when the sun went down at five o'clock or so, the Maules sat in the living room of their house huddled around the stove, trying to stay warm. If someone had to leave the room for some reason, then that person took the kerosene lamp with him so he wouldn't trip in the dark. Of course, that meant everyone else was left in the dark until that person returned with the lamp.

"There won't be any Christmas here," Walter Maule said. "We'll be lucky to eat."

Each family in Shallmar had its own story. Some Paul knew about, but others he only suspected or found out about later. While charity was necessary now, it wouldn't and couldn't last. Shallmar needed jobs either from the coal company or some other business. Neither seemed likely at the time.

13

Word Gets Out

To get a picture of how things were happening in December in the Maryland mountains, you should imagine Paul standing at the top of Backbone Mountain with a snowball. When he wrote those letters and made the telephone calls in late November, he made his snowball. When he talked to the reporter from the *Oakland Republican*, he let go of the snowball.

Well, that little snowball rolled down Backbone Mountain, faster and faster, getting bigger all the time and heading for unsuspecting Shallmar. It was on December eighth that people started noticing that snowball and their reactions added to the snowball's size.

The Republican had started publishing in March of 1877. Benjamin Sincell purchased it in 1890 and it has remained a family weekly newspaper since then serving Garrett County. In 1949, newspapers were still the main way that people found out what was going on in a community. Even so, *The Republican* was only published weekly. Garrett County was a rural community so there just wasn't much need for a daily newspaper. So by Thursday when *The Republican* was published, people looked forward to it. After all, they had a week with no news except for a few things on the radio. So they read the newspaper; all of it.

On Thursday, December 8, residents picked up their copies of *The Republican* to read: "Shallmar Residents Are Near Starvation, Urgent Appeal Made For Food, Clothing and Cash." It was a front page story under the masthead of the newspaper.

Mine closings and poverty were nothing new to the region, but the fact that it was so bad that children were fainting from lack of food and others not able to attend school because they didn't have warm clothing was more than anyone with a conscience could handle, and like Paul, it stirred them to action.

Charles Briner, the county director of employment security for Maryland, was inundated with telephone calls that spanned the gamut from pleas for him to do something to help Shallmar to accusations that he was killing the miners.

He had received a letter from Paul as well. Besides contacting his boss in the state government, he had also contacted the county's federal representatives about the possibility of getting surplus food for the county's many needy families. The government purchased food from farmers to maintain commodity prices. Much of this food then wound up being redistributed to the needy. U.S. Congressman J. Glenn Beall, who represented Garrett County, contacted U.S. Secretary of Agriculture Charles Brannon about getting surplus food for the county.

Briner's bosses in the state government chose not to work towards helping Shallmar at first. Instead, they started an investigation of the town. The Maryland State Police, at the request of the Maryland legislature, sent Corporal Thomas Currie to investigate the situation at Shallmar and report back about it. When Currie arrived in Shallmar and saw the conditions in town, he somberly shook his head. His report only confirmed that the newspaper hadn't lied.

Shallmar was starving.

Briner tried to defend the lack of immediate help for Shallmar. He told the newspaper reporters who suddenly wanted to interview him that 785 county residents were looking for work and the county unemployment rate was five percent or twice as much as it had been in 1948. Most of those out-of-work residents were coal miners.

"At least double that number are out of work," Briner said. "Many who do have jobs are only part-time." In Shallmar, the unemployment rate was probably eighty to ninety percent. Only nine people in Shallmar had anything close to a steady income: Paul, Baxter Kimble, four others, and three veterans who drew pensions.

Three other men were in Cumberland and Pittsburgh attending training schools. When they finished their training, it was expected that they would have to move out of town to be near wherever their new jobs were located. Until then, they were still considered Wolf Den Coal Corporation employees. They could live in company housing owned by a company that was not trying to collect the rent.

"There is no place for these men to turn," Briner said. "There has been no work since March and there was very little in 1948."

The problem had begun long before the newspaper article was published and was only reaching a head in December 1949. Shallmar miners had worked only three months in 1948 and twelve days in 1949. The reduced work in 1948 led to reduced unemployment checks in 1949. With such little work in 1949, they not only got very little pay, but they were no longer eligible for unemployment compensation in 1950.

The Wolf Den Coal Corporations had paid each of its employees "vacation pay" of $80 in June. It was actually more like a severance package since the mining company had already been closed for more than two months at that time.

A few of the miners with cars had found work in West Virginia, but their jobs were ninety minutes away. They would eventually have to move if they hoped to stay at their new jobs. Around a dozen of the miners had found temporary work on the Maryland state road crews or as farm help, but that work had ended in November.

Albert Males realized that the union he represented couldn't even protect his job. Albert, his brother and nephew had all found work during the summer picking apples on a farm. They had made as much as twelve dollars a day, though they still had to pay for room and board out of their earnings. It was well worth the cost since they didn't have to worry about being hungry. Anything left over was sent back to Shallmar for their families.

You might think that having worked for so long in the dark and under tons of rock that they would have appreciated being able to work in the sunlight with a worry that nothing heavier than the inspiration of Newton would fall on their heads. Not so. Working on the ladders to reach the highest branches of the trees had scared these tough miners.

Briner blamed part of Shallmar's problem on the coal itself.

"It would be hard to say exactly why, but it is a fact that coal from this mine is of inferior quality and unsuited for many industries since it has a high ash content," Briner said. "I have also been told by some operators that strip-miners corraled many orders during the strike last summer. That may partly account for Wolf Den's lack of orders." Coal quality was something that would come up more than once.

The United Mine Workers came under some condemnation for the situation, though such complaints only came from miners if their mouths had been lubricated a bit with a beer or something stronger. People in the coal business but who were non-union were more forthcoming with their opinions of the UMW.

While the union made it easy for miners to make a good living and hold onto their jobs, it made it harder for them to find jobs because seniority trumped ability. Even if other mines in the region had had openings, which wasn't happening because of short work weeks and fewer orders, the mines weren't likely to hire a Shallmar miner. Because of UMW-required health and retirement benefits that coal companies had to pay, many companies wouldn't hire miners over forty-five years old, which a lot of Shallmar miners were. Why hire a middle-aged man when you could hire a younger one who had a lot more work years left and there were plenty of them in Garrett County? Truth be told, a lot of small coal mining towns in the county were in just as bad as shape as Shallmar or pretty close to it.

"Lots of us are over forty-five—too old to get another job," Olen "Pap" Amtower, an out-of-work Shallmar digger, told a reporter.

The union had also made it very hard for any miner to work in 1949 in particular. The United Mine Workers had called four strikes during the year with the last not reaching any resolution until the beginning of December. That strike left miners working only a three-day week with a corresponding cut in pay. They went from earning an average of $200 a week to $60.

The Shallmar diggers had received unemployment benefits of $8 to $25 a week when they had received them.

Most diggers figured that with all of the striking that rather than digging coal, they were digging themselves deeper in debt. Most of them were looking forward to earning at least $1,200 a year less in 1949 than compared to 1948, which had been considered a bad year.

Their new lower wages came right at Christmas time, too. As one miner told a reporter, "It's just no good. We'll just earn enough to keep up with the grocery bills. Maybe we'll have a little left for Christmas." Not that the miner wanted his name used. Facing disappointed kids at Christmas was better than facing a ticked off president of his union local.

That digger's family could at least feed itself on three-day-a-week pay. Shallmar's diggers were getting no pay at all.

The Oakland American Legion Auxiliary was quick to announce that it was starting a collection of clothes and food, which would then forwarded to the Kitzmiller Lions Club. Paul had spoken with the Lions Club officers and they had agreed to help coordinate the distribution of the collected items. Members A. S. Barrick, Walter Sollars, Baxter Kimble, Harold Adams, Charles McIntyre and Wilbur Myers formed the committee to handle the Lions Club distribution in

Shallmar. Since Baxter ran the Shallmar company store, he would have a location from where he could handle giving the collections out to residents if it was needed.

Allen Weatherholt from the *Cumberland Evening Times* arrived in Shallmar on the day *The Republican* article came out. He interviewed residents for his own article, which ran the following day. The newspaper story in *The Republican* was also seen by reporters from other newspapers in the region. They wrote up their own stories and told their readers more information about the destitution in Shallmar. The newspapers with circulation areas closest to Shallmar were the first to run the stories.

Even if Paul had known all of the hullaballoo that the story would cause, while grateful, he would have frowned at how unresponsive the state was to the problem. Government policies could help or hurt the coal-mining business. A government agency controlled the availability of surplus food. Yet, the government officials were doing the most talking with the least results.

The diggers needed jobs. The union should have been trying to help them find work, but their hard-nosed resistance at the negotiating table was making it more costly to the mining companies at a time when demand for coal was low.

As Shallmar's story spread, newspaper after newspaper picked it up and asked, "How could this be happening in the greatest nation on earth?" Letters began filling Paul's mail slot at the company store. Each day, it seemed a few more letters were in the slot until finally all Paul was getting was a note from Baxter saying to ask him for the mail.

The other person who started getting calls and letters was Howard Marshall. Reporters tracked him down in Cumberland's Allegany Hospital recovering from what was said to be minor surgery. At fifty-nine, he wasn't in the best health, though, and he would soon retire from the coal business. He had been a patient in the hospital since the Monday before the first newspaper article was published. The reporters wanted to know why the mine wasn't open and what was being done about it.

He told reporters that he didn't know when the mine would reopen, "but I hope it will be soon."

I'm sure he did. Not only was he not earning any money with the mine closed, but he had lived in Shallmar for most of his adult life.

Howard was reluctant to talk about the problem since it would be airing company business.

"I can't quit. Mining's a gamble," he said somewhat cryptically.

He never fully explained the comment. I guess that he meant that there were good and bad times with the mining lifestyle. Just like gamblers can be flush or busted, so could miners. You could see this in the way that Dolly Crouse acted with her husband's pay; spending it on frivolous items when she could and knowing how to pinch pennies when there were limited funds.

At its peak in 1929, more than 179 tons had been dug from the Wolf Den Mine and 100 miners employed. Twenty years later, the mining company was virtually out of business.

He seemed to have little sympathy for the plight of his miners and their families. "I ain't seen anyone starving yet," Howard told the reporters. His solution was that the county welfare system should take care of them. "They pay enough taxes," he said.

He could have supported more than a few of his employees' families, but he didn't believe in charity, especially when he was paying taxes so that the government could provide it.

However, he was sympathetic to the miners' plight even if it might have been at Jesse Walker's urging. The company wasn't trying to collect on its house rent or company store accounts. At the company store, this created a secondary problem. It was hard to buy new stock to replace what had been sold when no money was coming in to pay for new stock. While the rent for the largest houses in town was only $12.60 a month, in some cases, rent hadn't been paid for over a year.

Despite the fact that Howard believed that there was at least fifty years of coal left in the mine, he took a dim view on the future of small mining companies like his.

With Howard in the hospital, Jesse was acting as the mine superintendent. He agreed with Howard, saying, "There is plenty of coal back in the hills, but it is difficult to get it out and put it on the market at a price sufficient to take care of the cost of operations."

Larger mines that were more mechanized could produce coal more cheaply. It wasn't possible with Wolf Den coal. It had to be hand mined for the most part in order to separate coal from rock.

Many people considered traditional mining to be fading, a thing of the past like buggy whip manufacturers. Yet just as there was still a need for buggy whip manufacturers for a much smaller market, mining would never be fully mechanized. Jobs within the industry would shift as business hopefully grew and in the end the goal was that the industry would be larger and employ more. The question was whether miners could adapt to the changing job market coming.

While Howard was hoping all of the commotion about the Wolf Den Mine would die down, that horse was out of the barn. Shutting the door wouldn't do any good.

The Shallmar powerhouse where the motors that kept the Wolf Den Mine ventilated and Shallar supplied with electricity were located. Photo courtesy of Western Maryland's Historical Library.

14

Help

Perhaps the biggest surprise came just two days after *The Republican* article ran. A large truck rolled into town and stopped at the company store on Saturday evening. From there, it drove out to the school. As soon as the truck left the store, Baxter sent someone to fetch Paul and have him meet it at the school.

Paul was surprised to see the truck pull up to the school because it was from the *Cumberland Times* and *Cumberland News* newspapers. For a moment, he thought it might be another reporter, but then he realized that a reporter wouldn't have come to town in a truck.

The driver climbed out of the truck and walked to the back to open the trailer door. Inside, Paul saw bags upon bags of food and clothing. Fresh vegetables, meat packed on ice, canned goods, milk, dresses, pants and shirts. All of the food looked fresher than anything at the company store even when it had had electricity to run the refrigerator.

"Where did all this come from?" he asked.

"People have been dropping it off at our offices since we had a story about the town in the papers yesterday," the driver told him. "They wanted to help, and it won't be the last load."

Since the article about Shallmar had run in the *Cumberland Evening Times* on Friday, people had been calling the office asking whether there was a central collection location for food and clothing for the town. So many calls had come in that the decision had been made just to tell callers to bring their donations to the newspaper offices. Other people called to say that they would drive their vehicles to Shallmar to deliver items and they needed to know who to deliver their donations to once they got to town.

Paul and one of the miners who had wandered over to the school to see what was happening unloaded the supplies into the newly painted union hall. Despite the below-freezing temperatures, other people began gathering at the school as word of the truck's arrival spread through town.

Seven-year-old sandy-haired Bob Hartman was among the crowd.

He watched through the window of the school as the piles of food and clothing grew taller inside the union hall. He couldn't believe that there was so much. It was like having another store in town.

He walked over to where the boxes were being unloaded from the truck. His eyes bugged out. Then he saw a set of new Levi overalls that looked like they would fit someone his size.

He told one of the men, "I'd sure like to have them overalls."

The truck driver looked at Bob and his threadbare clothes. "We'll see if we can't get them for you."

The man walked away. When he came back a minute later, he had the overalls in his arms and handed them to Bob. The young boy stared at them in disbelief for a moment.

"I can have them?" Bob asked.

The driver smiled broadly. "They're yours."

"Thank you," Bob said as he turned and ran for his house. Along the way, he muttered to himself, "I hope they fit. I hope they fit."

He tried them on as soon as he got home, pulling the stiff material over his worn pants and shirt. The overalls weren't a perfect fit, but it was good enough. He could roll up the pants cuff. That would be no problem. More importantly, he felt warmer without any drafts whipping through the holes in his pants like they had been. Though Christmas was still two weeks away, Bob felt like it was already Christmas morning.

Paul also marveled at the bounty that was growing before him and more would be coming the driver had told him. It couldn't have come at a better time. The first snow of the season had fallen a few days earlier and a thin coating was still on the ground.

How should it be divided so that it was fair for everyone in town?

With much of the town already watching the unloading, Paul called for a community meeting in the school to decide how to distribute the food. People crowded into one of the classrooms and listened to Paul explain that they needed to come up with a way to distribute everything because more would be coming. Some people seemed dumbfounded at the news.

The group formed a relief committee with Paul as the chairman, Martha as the treasurer, and members Frank Crouse, Kathleen Lyons, John Crouse and Gladys Hanlin. They made a list of the delivered items and started giving out food and clothing that evening. It was no use letting the food sit around when people were starving. Everyone in need would get something, but the larger families and the families with a greater need would receive larger distributions.

The committee got a surprise the next day when another truck arrived with donations for the town. The Frostburg American Legion had sent part of its collections for the area's needy to Shallmar. Cars also slowly drove up the street through most of the day, drawn by an editorial in the *Cumberland Sunday News*.

Boys at Shallmar School carry one of the early loads of food and other items donated to the town into the school. From the author's collection.

Among these cars was one that carried Allegany High School seniors, William Doub and Ross Keller. The two teens from Cumberland had spent the weekend making a collection of their own for Shallmar. Then they had loaded the food into their car on Sunday afternoon and driven it to coal camp 45 miles away. A group of Fort Hill High School boys from Cumberland had done something similar and raised $250, which they used to buy food and deliver it to Shallmar on Sunday.

Other cars would arrive in the coming days. Occasionally, they would hold people who just wanted to stare at the town they had read about, but most of the cars that arrived were driven by people who wanted to bring food, clothing or toys to help out the residents.

These people and others like them understood something that most of the politicians and bureaucrats did not. Promises never filled an empty belly. If the citizens had waited for the state or federal government to help, the people of Shallmar might very well have starved. For instance, though Shallmar would eventually get the surplus food that Paul hoped for, it wouldn't be until the following year. If he had waited for the federal or state governments to help, someone might have died.

That Sunday it looked like Christmas had come early to the town. Unshaven miners smiled behind their whiskers, mothers and wives laughed as children grabbed at the clothing separated into piles on tables in the union hall. Finding something they liked, many children hurried home to try on the clothes. Others couldn't wait that long and began pulling on sweaters over their summer shirts and trying on shoes. It was the first time in weeks that some of them had been warm.

Each family also got enough food—milk, fresh fruit and vegetables, canned goods and basic cooking ingredients—to last them most of the week. Some of the wives would probably stretch it out even longer, unsure as they were as to how long this bounty would last.

Two more trucks from Cumberland arrived midday on Monday, December twelfth. Paul let the students out for an early dismissal so that he and Albert Males could unload the trucks. The union hall filled faster than everything could be given away.

Some items seemed more popular than others. "I know of one ham that was stolen six times," George Brady said. "Someone took it from one of the trucks and hid it. Someone found where it was hidden and took it. Then someone else found it and took it and hid it and so on."

Seeing the piles of food in the union hall, Paul guessed that the families in town had enough food for two to three weeks, which turned out to be overly optimistic.

The Lions and Optimists clubs had set up a collection center at the rear of the A&P grocery store on Williams Street in Cumberland. Club representatives had also made arrangements to dispatch a truck to Shallmar daily with any donations that had been collected.

Paul met with Major Elmer Wall with the local Salvation Army and Lewis Ort of the local advisory board on Monday. They wanted to know how the Salvation Army might be able to help Shallmar. Paul said the residents needed basic food items that would last and could be used to help make meals. The Salvation Army agreed to donate flour, sugar and salt pork to the town.

Relief efforts for the town got a big boost when CBS broadcaster Edward R. Murrow saw the story on the news wires and told the country about Shallmar on his December thirteenth broadcast. Murrow had been a staple of radio news since he made broadcasts during the Battle of Britain in 1939. He wasn't afraid to allow his personal feelings and ethics to show in his stories. By 1949, he was the most-popular newsman on the air and the Shallmar story seemed tailor-made for him to express his opinions and have a lot of people hear it. He didn't make a direct appeal for donations, but he did say that Paul was accepting donations on behalf of the townspeople.

Along with food and clothing, more reporters, including Murray Kempton from the *New York Post*, started arriving in town to follow-up on the story of the town on the verge of starvation. As they poked around, other stories of deprivation began surfacing.

George Stonebreaker had been a miner for thirty-seven years. When the mine closed, he and his wife had needed to move from a company-owned house to save on rent. The new house wasn't much more than a shack, but it was off of company land. That didn't end their financial troubles, though.

"But I been sick and haven't been able to do anything with it," Stonebreaker said. "My woman worked a little bit, but now she's off."

They needed food and clothing, too.

Tony Crosco had been working in the coal mines since he was nine years old. Now he was a fifty-nine-year-old digger with long white hair and a family he couldn't support.

"I went in with a little pick and worked right beside my father," he said. "Later they put me to driving the mule, and I drove mules until I was seventeen. Then I started digging coal."

Despite his seniority in the mines, he was now just as out of work as the other Shallmar miners and he had no clue what he would do.

"I don't know," he said. "When us fellows get so old, we can't do much else. There's nothing else we can do. My, I sure hope we get back to work soon."

A week after the original *Oakland Republican* article had run, seven trucks had arrived in Shallmar with supplies and more than eighty letters with contributions totaling $500 had been received. With the arrival of the supplies and Paul's belief that it would carry the town through the rest of the year, many people thought that Shallmar's story had come to an end, but it was only the beginning.

15

Playing Political Football

By December thirteenth, just as Shallmar's story was first gaining significant national attention, the blame game was in full force with Shallmar being tossed around as a political grenade that no one wanted to be left holding.

Maryland State Budget Director James Rennie was the person who was holding the Shallmar grenade when it exploded. He found himself accused of violating the state's policy on buying coal from Maryland companies. The Maryland government purchased most of its 70,000 tons of annual coal purchases for state buildings and the Fifth Regimental Armory from the Banner No. 1 mine operated by the Stanley Coal Company in Crellin, Maryland. However, the rumor started circulating that the mine was actually in West Virginia.

The man who first made the accusation publicly was David Watkins, a UMW representative. This is not too surprising when you realize that the Stanley Coal Company used non-union diggers.

Peppered with questions from reporters and politicians, Rennie agreed that the Banner No.1 mine was physically not in Maryland, but he added that many of the miners lived in Maryland. They paid Maryland income taxes. They paid Maryland sales taxes on the goods they bought and real estate taxes on their Maryland properties. The Stanley Coal Company also used a Maryland address for business transactions, which meant that at least some corporate taxes were paid to Maryland.

"These men are 100 percent Maryland citizens," Rennie said.

He also pointed out that Stanley Coal Company's coal was of higher quality and at a lower cost than other Maryland mines could offer. This is because the non-union workers earned less than union miners, which allowed Stanley Coal to sell coal at a lower cost and still have the same profit margin as a union mine. The Stanley Coal Company also had the ability to fulfill its contracts and meet the state's needs, which is something not all Maryland mines could do. Again, this came from the fact that the mine was non-union and so when the UMW called a strike, Stanley Coal Company kept operating while other mines shut down.

Rennie told the members of the General Assembly that the state did occasionally purchase coal from the Krag Coal Company's Ream Mine in Preston County. It was only done in emergency cases, though even this coal was still dug by Marylanders.

The UMW jumped on this last bit and said that even this minor change was considered unacceptable when towns like Shallmar were in such dire circumstances.

Rennie then explained that the Wolf Den Coal Corporation had submitted a bid to supply the state with coal, but the mine had been closed by the time the contracts were awarded. The news wouldn't have been promising anyway. Wolf Den Coal Corporation officers said the company needed contracts to mine 5,000 tons of coal a month to justify running the mine, but at best, the company might have gotten only 500 to 1,000 tons of coal a month in state contracts.

"Our policy is to try and not put all our eggs in one basket," Rennie said.

If this had been the company's only contract, it would have been able to operate the mine for a month, if that.

Even with the small contract, the company may not have been able to reopen. Not only were there the safety improvements that the state mine inspector had recommended that still needed to be put in place, but Howard Marshall was telling people that the mine also needed at least $100,000 in new washing machinery to improve the competitive position of the Wolf Den Coal Corporation. He doubted that the mine would be able to get it and without it, the mining company would only be able to produce coal for the spot market to sell for immediate delivery. A coal company couldn't run without the stability that long-term orders provided.

Rather than looking to mining for future employment, Howard said the best chance for miners to get local work was to get hired on the flood control project on the North Branch Potomac once it got started. This had finally come about to protect the riverside communities from the damage that flooding like Shallmar had experienced in 1924.

Paul ignored the politics and focused on the children. He told the reporters, "Even if these miners return to work—and the chances don't look too good now—it will be some time before the children get the diets they need."

Albert Males told reporters that small mines like Shallmar hadn't done well since Congress had passed the Guffey Act in 1935. Democratic Senator Joseph Guffey pushed the Bituminous Coal Conservation Act and it came to be known by his name. The law was part of President Franklin D. Roosevelt's New Deal. It set coal prices to protect small mines from being undercut by larger mines, which could underbid them. The U.S. Supreme Court ruled it unconstitutional and said that the government couldn't control coal prices because it infringed upon free enterprise. The law was adjusted and the Guffey-Vinson Coal Act was passed in 1937, which was found to be Constitutional but still accomplished the same thing and allowed the government to control prices.

The problem that politicians hadn't foreseen was that purchasers could buy high-quality coal for the same price as lower-quality coal. Why buy a Ford when you could buy a Cadillac for the same price? Coal buyers chose the high-grade coal. Many times this left Shallmar and other mines with lower-grade coal without a business.

The UMW ran full-page ads in the Cumberland newspapers with the headline: "Do we want any more Shallmars?" The copy urged readers to write to Maryland Governor William Preston Lane, Jr. and members of the legislature urging the state to only buy coal from Maryland businesses, which for the most part, were all unionized.

Maryland State Delegate Robert Kimble appeared to have been someone who was listening to the people who responded to the ad and the call for action. Kimble was a Republican who represented Allegany County in the Maryland Legislature and he was also the minority leader in the legislature so he had some power in the government.

Governor Lane had called a special legislative session for Saturday, December seventeenth, specifically to deal with a $50 million school bond issue that Maryland Court of Appeals had ruled was invalid because of a clerical error. The governor wanted the session to deal with the bond issue and a few other non-controversial matters with the hopes that everything could be taken care of quickly.

As the time for the special session drew near, Kimble announced that he planned to introduce some relief bills during the session in response to the situation at Shallmar. He wanted to use money from the state surplus to increase unemployment payments and welfare payments to families of unemployed.

This didn't sit well among others, including leaders of his own party. Senator Stanford Hoff, the chairman of the Republican Party in

Maryland, said, "It is not a party matter. It's strictly a private affair with Kimble. He has never brought it up in party caucuses and the party is not prepared to support or oppose it. As far as I am concerned, I am against it at this time."

The governor, who had hoped to keep the special session to just a few hours on Saturday morning, now saw the possibility of it stretching out for a day or two.

On top of this, two delegates from Baltimore's third district, J. Raymond Buffington and Chester Tawney, asked the delegates to forego reimbursement of their travel expenses for the special session to which they were entitled. If the senators and legislators voted to be paid their expenses, then Buffington and Tawney planned on asking them to donate the expense money to the families of Shallmar.

John T. Jones, president of the UMW District 16, sent a telegram to Governor Lane that used Shallmar as the reason that the state should be buying only from Maryland companies. Part of it read, "No doubt you are familiar by now with the terrible condition now existing at Shallmar, Md., due to the Wolf Den Coal Corporation mine being closed down since March 1949 for lack of orders for coal.

"Suffering now being endured by the coal miners and their families could and would be eliminated by the State of Maryland giving the Wolf Den Coal Company some of the coal tonnage consumed by state institutions."

Jones said that by buying coal in state, even if it was more expensive, Maryland would save money that it was spending on relief payments to miners.

Following the public release of the telegram, A. G. Uhl with the Maryland Department of Budget and Procurement, told the legislature that none of the coal that the state purchased could be considered as coming from "outside sources." Stanley Coal Company was a Maryland business and the majority of is employees were state residents.

Yet, while the politicians were quick to use Shallmar to achieve their goals, they were slow on delivering on their promises to help the town and its residents.

16

Children Who Like School Lunches

One of the early efforts to help Shallmar took a little bit longer to get moving, but given that it was a government action, it was probably pretty quick for them. It's kind of like saying a turtle is fast, but only if you're a sleepy rabbit who just lost a race.

By the time the *Oakland Republican* article had come out, the Garrett County Commissioners had already decided to fund a hot lunch program for Shallmar School, a building without a kitchen or cafeteria.

The idea of serving hot lunches to students to students at school rather than having them bring a packed lunch to school or go home for lunch was not a new idea. In fact, a coal mining town in neighboring Allegany County had been a pioneer of the idea in 1939.

In August 1944, the Allegany County Board of Education requested a part-time nutritionist from the Cumberland and Allegheny Gas Company. Nutritionist Flora Dowler was sent to assist the board and W.P. Cooper, the food program director for the board, who was already a pioneer in school food service.

Dowler's work began with two days of training for the cooks at Fort Hill High School in Cumberland where she demonstrated and prepared different types of lunches that met state guidelines and could be prepared at the school. She followed up that initial training with similar visits to each school at least once a month.

During those visits, she discovered a lunch program that had been started in 1939 in Hammond Street Elementary School in Westernport. The school board, working with the Works Progress Administration, the Surplus Marketing Corporation, Maryland Department of Public Welfare and the school's Parent Teacher Association provided free lunches to sixty percent of the students at the school. Dowler said this was needed because "most of the bus children attending this school came from the region where the coal mines were practically abandoned and the parents unemployed."

The program's success had led to its expansion to all other schools in the county.

Paul hoped to get a similar program into Garrett County schools, as well. The War Food Administration and the Maryland Department of Public Health supported the Allegany County program and allowed it to operate on a countywide basis. The Maryland Department of Public Health regularly examined the seventy-six cafeteria employees and inspected the lunch rooms to make sure they were sanitary.

The school lunch program had expanded countywide by 1945. The cost of a school lunch was sixty cents a week or fifteen cents daily. More than two million lunches were served in Allegany County schools in 1945.

The idea of having such a program in Shallmar School intrigued Paul and he worked with the Garrett County Board of Education and the county commissioners to figure out how it could be done locally.

The students at Shallmar School couldn't afford the sixty cents a week, but if Paul could get the hot lunch program into his school, then he would find a way to get the money that would be needed to allow the students to get a free hot lunch.

Finding a place to eat was the easiest problem to solve. With the coal mine shut down, the union hall at the back of the school wasn't being used. It was a large enough space for everyone to eat together once tables were set up. The union already had chairs stored in the room for when they did meet.

The more-pressing issue was how to prepare hot meals at the school. The Garrett County commissioners weren't willing to pay for an expansion at a school that they had pretty much decided to close and even so, it wouldn't have been ready until the new year at the earliest. The kitchens in the houses in Shallmar were too small to prepare hot lunches for large groups.

Then someone realized that Kitzmiller School was a hop, skip and a jump away from Shallmar by car and the school had a hot lunch program cooked in the school's kitchen. The lunches for Shallmar School could be cooked at Kitzmiller School, dished out on plates that could be covered and driven to Shallmar to be served while they were still hot.

With a plan in place, the commissioners allocated $1,200 to feed the students at Shallmar through the end of the school year. The unemployed miners at Shallmar got together and painted the walls of the union hall to give it a fresh look in preparation for its use as the school's new cafeteria. It also got rid of the dirt and smoke stains from the miners who used the room as a union hall.

On December seventeenth, the students at the school sat down at two long wooden tables and had their first hot lunch in weeks, if not months. School lunches at the time included things like roast beef, mashed potatoes and green beans. You can be sure that the kids in Shallmar School had no qualms about eating anything they were served. They were full for the first time in weeks.

Students at Shallmar School get their first full meal in weeks as the school lunch program starts. Photo from the author's collection.

17

Plenty

For most people in Shallmar, the politics didn't mean anything even if they knew about them, which was doubtful. No one had telephones. They couldn't afford to take a newspaper and radio reception in the shadow of a mountain was spotty at best.

Outside the area, word was spreading, though. Paul and others in town realized that newspapers must have picked up on the story because of the postmarks on the letters that were coming into the post office. By the middle of December, eighty letters and $500 had been received to help Shallmar's residents.

And the numbers kept increasing.

"I have answered at least 100 letters and together (with Mrs. Andrick) we have written 1,300 postcards in an effort to thank the people of America for what they have done for our community," Paul said near the end of the month.

Trucks were arriving on a daily basis and not only from Western Maryland. Trucks braved the steep, narrow roads to reach Shallmar. A Baltimore meat packing company sent the town a load of fresh beef. The Maryland Jewish War Veterans began collecting toys for the children of Shallmar. The Amici Corporation in Baltimore sent fifty dollars to the Oakland Chamber of Commerce to purchase candy for the children of Shallmar. Plus, Amici employees raised another $1,000 for the town in general.

The American Legion in Baltimore began collecting food and clothing to the extent that it filled a room at the War Memorial Building in the city.

"I have had to clear a path to my desk," State Adjutant Daniel H. Burkhardt told the *Baltimore Sun*.

Among the donated items for Shallmar was 300 pounds of bacon, 250 loaves of bread, 200 quarts of milk, cases upon cases of canned goods and groceries, 100 pairs of shoes, twenty men's overcoats, twelve women's fur coats and blankets, not to mention toys for the children.

"If I couldn't see it with my own eyes I wouldn't believe such a response was possible in just three days," Burkhardt said.

The items were loaded onto three large delivery trucks on December sixteenth that were usually used by a Baltimore exterminating company. Joseph Blotkamp owned the trucks and donated them for the American Legion to use and deliver on December seventeenth.

Students decorate the Christmas tree at the school. Photo courtesy of Jerry Andrick.

From New York City, the Save the Children Federation said that it would be sending toys and other needed things to the town. The Cumberland local of the United Brewery workers of America also began raising money for the town residents. Cumberland dairies were donating milk and bakeries were donating bread to Shallmar.

Much of the food and adult clothing was distributed right away, but the other items were kept in the Andricks' home and began accumulating. The Shallmar Relief Committee had come up with an idea for how it would distribute toys and other items it received.

The story of Shallmar's plight even made international news. Baxter Kimble sent a clerk to fetch Paul to the company store on December seventeenth because he had a phone call. When Paul picked up the receiver, he found himself speaking with a reporter from the *London Times*.

Letters also arrived from other European cities like Paris with words of concern for Shallmar's residents.

Some of the photographs taken in town became widely published in December.

During the distributions that had taken place during the weekend after the first article about Shallmar ran, a photographer took a shot of two-year-old Jean Ann Crosco sitting amid a pile of shoes and crying because none of them fit her. Now Jean Ann was a happy child generally who most people could find sitting in a cardboard box and playing quietly. On this particular day, Georgia Crosco, Jean Ann's mother, watched someone actually pinch her daughter to get her to cry at just the right time.

Numerous newspapers all around the world ran the picture and donations began pouring in specifically for Jean Ann. About a week after the photo started running, Georgia told a reporter, "We have heard from every State in just a week. The kind people sent her dresses and other pretties. And she had about $200 in money gifts. And Jean Ann has forty—yes, forty—pairs of shoes."

The picture also spurred A. K. Rieger, the president of Gunther Brewing Company in Baltimore, to buy all of the children in Shallmar a new pair of shoes. Between that free pair and other donated pairs, some children wound up with two pair of shoes. Dolores Crosco, whose mother had kept her out of school because she had no shoes to wear, got two pair of new shoes, one of which was girl's cowboy boots. Her brother, Thomas got a new pair and a battered pair of white summer shoes, which was still better than anything he had.

With all this help pouring into town, it would have been easy for the residents to stop trying to take care of themselves, but that wasn't how miners lived. The Kitzmiller Boy Scout Troop, to which some of the boys of Shallmar belonged, began collecting broken and worn out toys from the area. They went door to door asking for donations, looked in trash bins and searched their own closets for old, broken and worn-out toys. Robert Young, a teacher and the scoutmaster, coordinated efforts among the boys and men in the area to recondition the toys in order to give a child a like-new toy on Christmas Day.

And it was turning out to be a very merry Christmas for the children of Shallmar that year.

On December twenty-first, the senior class at Ursuline Academy, a Catholic girls' school in Cumberland, cancelled its own Christmas party to throw one for the children of Shallmar in the Shallmar School. Two days later, the United Paper Workers Union Local 67 came to Shallmar to throw the children another party.

While the good news seemed to be showering down on Shallmar, some people were determined to seek out the negative news. *The Republican* newspaper printed a couple of unconfirmed rumors about Shallmar. One story was that some of the residents were trying to sell their donated goods in town across the river for cash so that they could buy beer and whiskey. This wasn't too surprising to a lot of people. Every town had its bad apples and more than one man in Shallmar enjoyed his booze too much. Another story was more surprising. A farmer delivering a load of potatoes to Shallmar had offered a resident three dollars a day to work on his farm and had been turned down. The man said he didn't need to work because of all the help the town was getting.

Even some Shallmar residents expressed doubt about their neighbors need for the country's generosity.

"They took advantage of it," Margaret Morris, Jesse's niece said. She believed that the situation looked worse than it was but that it was typical of a coal miner's life.

When Buelah Milavic saw everyone rushing for food and clothing, she said, "I wouldn't stoop to that level. They don't need it. They're taking charity."

These stories didn't get wide circulation, but it started some people wondering if things were as bad in Shallmar as they had been led to believe.

The Labor Herald, a Communist newspaper in New York City, sent a reporter, T. O. Morrow, to the county to view the situation. His article, which was reprinted in other newspapers, played up Shallmar's poverty as a way to try and make the case that capitalism didn't work.

However, the donations continued to arrive by truck, car, train and mail. The Cumberland Optimist Club alone had delivered more than fifteen tons of food and clothing to the town by the middle of December and it was only one organization of dozens that was delivering food.

Senator Beall continued to push for getting surplus food for the town, but he had to fight against the bureaucracy. The board of education was the local sponsoring authority that needed to request surplus food. And then, the food could only be used for school lunches.

Students decorate the Christmas tree at the school. Courtesy of Jerry Andrick.

18

A Jolly Christmas

As Christmas Day approached, the downstairs of the Andrick home continued to fill with boxes and bags of letters and postcards; most of them with cash and checks in them. The mail came in on the evening train that stopped in Harrison just long enough to drop off the day's mail. As postmaster, Baxter Kimble would usually walk across the swinging bridge to Harrison and pick up a satchel of mail for the entire town. This month, however, the volume of mail had been growing so steadily that he was driving a truck to Harrison to get the mail.

It seemed that for no matter how long the Andricks spent writing out "Thank You" postcards to everyone who sent a letter or package, the pile never seemed to grow any smaller.

On the Friday before Christmas, Baxter handled 2,000 pounds of mail that the railroad delivered to Harrison. On a normal day, that would have overwhelmed Baxter, but he had plenty of time on his hands because no one was buying anything at the store. They didn't need to. Everything had been donated to them this year.

A week before Christmas, the Shallmar Relief Committee decided that since the town's residents had been blessed by the kindness of so many people across the country that they should be just as willing to share their surplus with the same generosity.

Shallmar was the hardest hit of the coal-mining towns, according to Charles Briner. Still, it was far from the only town where miners where watching their jobs disappear. The demand for coal did not rise to the level of the supply. Diggers were working reduced hours or none at all.

By Paul's estimate, the town had received about $20,000 in gifts and more than $5,000 in cash and checks, which was probably more money than all the miners in town had made that year and perhaps in the previous two years combined. And this amount did not include anything that went directly to someone other than the relief committee. Members of the Shallmar Relief Committee would sort through the mail, answer each one and record any checks or cash. Molly and Paul would drive to Oakland every other day or so and deposit the money into the town's checking account that the committee had set up.

One of the first things that Paul had the committee do was to authorize paying the student portion of the hot lunch program for Shallmar's students.

The committee began sending out food and clothing from their stores to help another seventy families in the region who were in just as bad a shape as the families in Shallmar had been. Some got money, some got clothing and they all got food.

"Today with $5,000 on deposit to their credit in a nearby bank, with tons of canned food piled up in their schoolhouse and more toys for their children than they know what to do with, the people of Shallmar have a realization of the Christmas message of "good will to men," one reporter wrote.

On the Friday before Christmas, the members of the Shallmar Relief Committee sorted the toys, clothing and other items that had been held back from distribution so far and carried them into the Big Room and Little Room. Then they used decorations saved from the Christmas parties that had been used for the Christmas parties thrown for the children at the school and decorated both rooms.

While this was happening, parents arrived at the school and drew a piece of paper that had a time written on it from a bowl. This was the time they would be allowed to come to shop to select gifts for their children. The committee set things up this way to ensure that there wouldn't be a mass of parents packing the room, pushing and shoving to get the items that they wanted. The idea was for the parents to come to the toy shop by themselves so their children wouldn't see the gifts that their parents picked out. That way, they would be surprised on Christmas morning.

As each set of parents entered the school, they were allowed to get a book, a toy and a novelty item for each of their children. The room was filled with more than enough items to fill that demand twice over with additional items still arriving by truck.

The parents came and made their selections. Some of them made their choices looking at the gifts through tears in their eyes. Once chosen, they would hide the gifts in bags and head back home.

Christmas Eve day was for the children. They began arriving at the school and lining up an hour before the announced opening of the toy shop. This would be their time to select their own gifts for themselves. Though there were no adults to supervise them, they weren't wild and unruly. They waited patiently in the cold as the Big Room was restocked with items from the Union Hall and the Andricks' house.

When Paul appeared in the doorway of the school, they quieted down. He explained the rules of the day to them. Only five children at a time would be allowed in the Big Room and each child would have fifteen minutes to select their gifts. Each child would be handed a slip of paper with the time that they entered the toy shop on it. He also reminded them that this was still a school and he expected them to behave as such.

When the students entered the Big Room, they were faced with toys, new and refurbished, piled on top of and under tables. On the blackboard at the front of the room, Paul had written in large letters next to where he had written the Pledge of Allegiance, "For each child, one toy, one novelty and one book."

The children were stunned at the abundance that confronted them and were hesitant to pick anything up. The girls, especially, were slow in selecting their dolls. They would look at the dolls without touching them and once their decision was made, they would pick up their doll and hug it.

They would clutch their gifts and walk to the exit of the school. Once outside, their excitement could no longer be contained and they ran home.

Shirley Hanlin picked out a dress, a pair of gloves and a new doll.

Of all the children who came to the toy shop that day, only one returned. Six-year-old Elaine Paugh came back about an hour after she had finished her selections. She quietly told Paul that her mother had told her to return the very nice doll that she had selected for herself. She was obviously fighting back tears as she handed the doll to Paul.

Paul felt sorry for her as he told her to go ahead and select another doll, but he knew something that Elaine didn't. Her mother had picked out a beautiful doll for her on Friday evening and thought that another little girl should get the nice doll that Elaine had picked out. Elaine would be sad for only a day. Tomorrow, she would get the doll her mother had selected. For today, she would love the fifty-cent rag doll.

As the piles of toys and other gifts shrank, more were brought in from the Union Hall where Albert Males and some other men from the town unloaded the trucks that were arriving filled with food, clothing and gifts. Trucks came from Hagerstown, Frostburg, Accident, Oakland and Montgomery County from the Kensington View – College View Citizens' Association.

The Montgomery County truck had "Project Shallmar" written on the side of it. Seeing the truck upset Howard Marshall a bit since "Shallmar" came from his last name. It seemed somewhat like saying he couldn't take care of his employees.

That evening, Paul dressed up as Santa Claus to deliver bags of candy, nuts and oranges to each of the families in Shallmar.

Charlotte Crouse and her brother were fighting that evening and too excited about the coming of Christmas to go to bed. Her mother finally had enough and told them, "If you don't stop fighting, Santa Claus won't come."

At that moment, Charlotte looked up and saw Paul outside the window in the living room. Of course, she thought she was seeing Santa Claus. She quickly stopped fighting with her brother and was on her best behavior for the rest of the night.

The next morning Charlotte got the doll for Christmas that her parents had picked out for her on Friday.

"I loved that doll," Charlotte said. "I wouldn't let anyone touch it. It was plastic with long hair."

During the Christmas morning church services, the pastors and preachers announced that children from some of the surrounding towns would be welcome to come and pick out their own toys at the Shallmar School from 2 p.m. to 4 p.m. The relief committee had decided that it wanted to make sure that all of the gifts were shared with other children who might not have a Christmas otherwise.

"And the person who brought this about is one of the disappearing race of male country school teachers, J. Paul Andrick, who fought for his kids until he won," one reporter wrote about the events of the day.

19

A Slow Death

Donations to Shallmar and its residents continued flowing into town even after Christmas. With hunger only temporarily eased by what grew to an estimated $7,000 in cash and $30,000 in food, toys, clothing and other items by early 1950, Paul and the Shallmar Relief Committee turned their attention to long-term solutions for the town's problems. The money, which had been deposited in a bank in Oakland, lasted until the account was closed in May of 1952. At that time, an audit of expenses showed that $3,685.59 had gone to buying lunches for children, $1,513.27 had been direct relief to families and $944.40 had paid for health care, plus other expenses. Donations came from every U.S. state except for Utah.

With the beginning of a new year and a new decade, Howard Marshall sought out new contracts for coal so that the Wolf Den Mine could be reopened, but he was having limited success and it didn't look like it would be getting easier.

Maryland State Budget Director James Rennie stepped in it again in February with regards to Shallmar. He told the House of Delegates Ways and Means Committee that month that the Shallmar miners would need to find a different job rather than waiting for the Wolf Den Mine to reopen "or they are going to be on relief forever."

Shallmar came up during a budget discussion about what to do about workers who had exhausted their unemployment benefits.

"There's no answer for the Shallmar mining situation," Rennie said. "The coal is no good. It can't be sold in a competitive market."

The delegates were not happy with the answer since it didn't answer the immediate question about what to do. They decided to take action. They had budgeted $300,000 to help out "unemployed employables." The governor wanted the amount to be $500,000.

It was the first time in years that the state had offered any state help to unemployed workers who didn't meet any of the state's exemptions – physical handicaps, old age or women with small children, and it was Shallmar's plight that brought it about.

"We were afraid we were heading for a welfare state, and that's only one step from Communism," Delegate Leroy Pumphrey, a Democrat and chairman of the committee said.

"The governor can't see a man go hungry because he can't find work through no fault of his own," Rennie told the committee.

"Is there anything wrong with asking these people to do a little work for their money?" asked Delegate Horace Whitworth from Westernport.

Whitworth had introduced a bill that allowed counties and towns to employ relief recipients on any public works projects. Many representatives thought it was a good idea, but the idea came to naught. Rennie's comments didn't earn him any friends in Garrett County, though.

Leslie Sharpless, the president of Sharpless Coal and the mayor of Kitzmiller since 1934, was quick to defend Garrett County Coal in a letter to Rennie. He pointed out that half of the supply of coal that the state used came from the Crellin Mine, which was on the same coal seam as the Wolf Den Mine. So if the coal was fine for the state to use, how could Rennie say it was inferior?

Besides the Wolf Den Mine and the Crellin Mine, Johnstown Coal and Coke, McNitt Coal Company, Garrett Coal Corp. and Sharpless Coal Company were also on the Upper Kittanning Coal Seam and no one said their coal quality was inferior.

Sharpless told Rennie that his company had employed a dozen of the miners from Shallmar in the past month. "We're very much disturbed," the mayor wrote. "We've done our best to give these men employment but it doesn't look like we're getting much help from the state."

Howard Marshall's short visit to the hospital in December turned out to be a symptom of greater health issues that forced him to remove himself from business matters of the mine. Jesse Walker stepped in as the acting mine president. He believed that not only was Wolf Den Mine's coal quality acceptable, the mine had plenty of coal in it waiting to be mined.

The problem that Jesse had was finding the capital needed to reopen the Wolf Den Mine

With an energy that Howard hadn't been able to show recently, Jesse made plans to reopen the mine, but first he had to wait until yet another nationwide coal strike was settled.

Jesse managed to raise $30,000 in order to reopen the mine. With more than a little fanfare, the Wolf Den Mine reopened for two weeks at the beginning of March 1950 after the UMW strike ended. It seemed like it was a continuation of the Shallmar Christmas miracle, but just as the Christmas season gave way to the New Year's "blahs" of January and February, so the hopes for continued mine operations gave way to reality.

A skeleton crew of twenty miners was said to have mined 600 tons of coal for the spot market, though the official reports don't show it. After two weeks, the mine closed down again and did not reopen in 1950.

Beyond that, the prospects didn't look much better for it to ever open again. The Shallmar miners couldn't bet their lives on a large contract coming through, not while larger coal companies could undercut Wolf Den Coal Corporation's prices.

The Wolf Den Coal Corporation appeared to be the first domino to fall among the Western Maryland Coal Camps. One year to the day of the Wolf Den Mine closing, the Manor Mine No. 3 and strip mine at Vindex closed because there was no market for coal. The closure put 180 miners out of work.

A few months later, the Davis Coal and Coke Company Kempton Mine closed putting another 218 miners out of work.

The Garrett County Commissioners voted to ask the state government for help in April. On April twenty-fourth, they passed a resolution seeking help because "practically all of bituminous coal mines in this county have closed down, most of them permanently."

Among the many letters Paul had received about the situation in Shallmar, one was from Diane P. Gradman with the Melrose Canning Company in Hampstead, Maryland. The company had jobs it needed filled and a proposition for the miners in Shallmar and Vindex.

Gradman offered to pay the expenses for three representatives to visit the plant and view conditions there for themselves. If the committee found Melrose to be a company they could recommend, then Gradman would present his employment offer.

George Brady, his wife Melissa and Sam Tasker, Jr. traveled to Hampstead and toured the factory. They returned excited. The company was clean and reputable. It had been founded by Isadore Gradman after he emigrated from Lithuania in 1925. When Isadore died in 1940, he left the company to his second wife, Diane, who had been running it since that time.

Gradman offered to house the miners and their families while they worked for the company, which was expected to be at least until late October. The group recommended that the miners accept the offer.

Thirty families initially took jobs with Melrose Canning and another twenty followed a few weeks later. The families moved to Hampstead where they had guaranteed work through the fall. The wages they earned during that time would be more than they had earned in 1948, 1949 and so far in 1950 combined.

Though the work was above ground, it was no easier than coal mining and it had its own dangers. The cannery employees found themselves working sixteen-hour days to get the job done, according to George Brady. He worked at the cannery during the summer and fall of 1949 with his father while his mother stayed in Shallmar.

"We would shell peas for so long that we couldn't open our hands at the end of the day," he said.

As for the dangers, George was working outdoors on a large piece of equipment used to thrash. It was essentially a large vat that had I-beams that crossed the top of it. The thrashing arm hung down from the I-beam. One day a storm blew through and George warned two men that they had better get off the I-beams in case lightning hit one of them. He forgot the fact that he was also standing on top of the vat.

Lightning hit the metal and threw him into the air off the vat.

"Oh my God! It killed him!" George remembers his dad yelling.

George hadn't been killed, though the lightning had roughed him up a bit. His feet and butt had been burned slightly by the raw electricity and he was shaken up, but by and large, he was all right.

The move of families to Hampstead pretty much cut the population of Shallmar in half. Many of the younger residents, particularly those that graduated in June, moved out to other areas to find work. Once they did, they sent money home to their families.

Other families found work elsewhere. Charlotte Crouse's father went to work in Ohio and moved there to live with his brother. He returned to see his family only occasionally. They had moved to Kitzmiller and lived in an inexpensive half-house there.

A ripple effect from the Wolf Den Mine closure was that the Garrett County Board of Education decided to go with the school restructuring plan that included the closure of Shallmar School.

The parents of Shallmar, Vindex and Kitzmiller signed a petition to keep their children out of school when school began again in September unless the board of education agreed to make improvements to Kitzmiller High School and keep the high school in Kitzmiller. The petition also noted that the parents favored the construction of the new high school, but they opposed further consolidation. The submitted petition had 339 signatures on it. Kitzmiller School had seven classrooms, gymnasium, a fully equipped shop, cafeteria, home economics room and a library.

A group of parents met with the board in early September and was unable to reach an agreement. When Garrett County schools opened on September sixth, it was expected that around 300 students would attend Kitzmiller School, Shallmar School, Vindex School and West Vindex School. The school board had even rented two rooms from the Kitzmiller American Legion to handle what was expected to be an overcrowded Kitzmiller School.

Thirty-four students showed up the first day. This is not thirty-four at one school. That is the combined attendance at all four schools.

It was an act of civil disobedience on the part of the parents that not everyone appreciated.

"This looks like parents deliberately inviting their children to disobey law," *The Republican* wrote in an editorial. "It borders on taking law into their own hands the same as these roving bands of pickets who forcibly compel law abiding citizens to abandon their right to earn a living for their families."

Six men were eventually arrested for "conspiring to cause the children to strike." One of the men was the mayor of Kitzmiller and another was a doctor in town. Kenny Bray was also among the men indicted.

The judge set the bail at $5 each, but the town was too poor to raise the $25 so he let them go, saying, "Well, you certainly aren't going anywhere."

Even the indictments failed to entice students back to class. In fact, total attendance dropped below 20 students. Under state law, teachers were sent to students' homes to try and find out what the problem was even though everyone already knew the answer to that question.

When the case came to trial, the judge ordered the students back to schools so that the county wouldn't lose state funding because such a high number students who wouldn't have attended at least 180 days of school. As a compromise to the parents, the judge also ordered that certain improvements and repairs be made to the school.

This didn't solve the problem. It only delayed a resolution.

At the 1951 Kitzmiller High graduation – the final one – the county superintendent, R. Bowen Hardesty, was scheduled to be the commencement speaker. Given how bitter the fight against closure had been, no one in Kitzmiller or the surrounding towns wanted to see anyone from the board of education at the graduation. It would have been like a slap in the face for residents.

Some of them hung a dummy from a tree in front of the Methodist Church in Kitzmiller with a sign on it to say it was supposed to be Hardesty. Jenny Keller stood guard in front of the dummy to keep anyone from cutting it down.

When Kitzmiller School Principal Thomas Baucom went to find a telephone to call Hardesty and tell him not come because of the anger towards him in town, Baucom saw the hanging dummy.

"Jenny, you get that down right now," he ordered the woman.

Keller put her hands on her hips, threw back her shoulders and said, "You make me!"

From most other women, it might have been considered an idle threat, but Jenny Keller was as large and strong as any man.

Another one of Jesse's sisters, Caroline Wilson, was the superintendent of grammar schools for the county. She came into the church and sat down with her family. She heard some of the rumbling in the room and realized that she shouldn't be in the church since she worked for the board of education.

However, the commencement music started playing so she settled down with the other adults to watch as the students started marching in. Before they could be seated, Baucom was at the front of the chapel.

"Stop! There won't be a graduation this evening," he announced.

A low rumble started through the crowd. "Oh yes, there will be," someone shouted.

"No. There won't."

And with that, he took the diplomas and headed outside.

A parent stopped him in front of the church and tried to turn Baucom around and send him back into the church. The two men got into a loud argument until the parent finally said, "I don't hit a man with glasses."

Baucom took off his glasses, but before he could do anything more than that, the parent punched him in the face.

The next day Principal Baucom, his nose swollen and his face bruised, walked into the cafeteria.

"People think I was physically hurt," he said as he shook his head. Then he tapped his chest. "Here's where I was really hurt."

He then proceeded to hand out the diplomas to the seniors without any fanfare and without their families being able to attend.

Cancelling the graduation made Baucom a pariah in town. People wouldn't talk to him. When he came into the drug store where Margaret Walker worked and asked for a newspaper, she told her principal, "Mr. McIntyre [the store owner] told me I can't wait on you."

The final straw came one night when some of the men in Kitzmiller carried rocks from fields and the river and piled them in front of every door of Baucom's house so that he had to climb through a window to get out of his house.

Baucom moved out of the area after that and became a vice principal at Southern High School.

The following fall fewer than ten children boarded school buses to make the thirty-five mile round trip to the new Southern High School. All of the other 110 junior and high school students in the area chose instead to only travel a mile up the mountain in West Virginia to attend Elk Garden High School.

This was due to a temporary compromise that Maryland State Schools Superintendent Thomas Pullen had agreed to. The state would pay the tuition, books and transportation cost for any student in the Kitzmiller School who enrolled at Elk Garden High School. The response had been so overwhelming that the West Virginia school needed to build four additional classrooms onto the school.

Even the former teachers from Kitzmiller High wouldn't transfer to the new high school. They chose to get jobs with other school systems or Frostburg College.

Shallmar also made national headlines once more in the fall of 1951, though few people realized that they were reading about people from a town that had nearly starved two years earlier.

A group of around fifty miners from Shallmar, Kitzmiller and Vindex trained in mine safety practices as the parents held their children out of school. Since the mine at Shallmar was shut down, wooden tunnels were built in the town where the men were run through training exercises under the supervision of Harry Buckley, the district mine inspector; L. C. Hutson of the University of Maryland and Fred Baker of the U.S. Bureau of Mines.

The six best diggers were chosen to go to Columbus, Ohio, to compete in the National Mine Rescue and Safety Championships. It was the first national championship to be held in twenty years.

The competition was held at the state fairground on October second. The team was one of fifteen teams from Ohio, Illinois, Kentucky, Maryland, Pennsylvania and West Virginia competing for the national title among coal miners. The teams were sent into a constructed mine shaft to deal with a simulated mine disaster. It was a test of the miners' equipment, physical condition and ability.

The Kitzmiller Mine Rescue Team included men from Shallmar. Photo courtesy of Robert Hartman.

At the end of the competition, the team from Kitzmiller walked away with the national title. Carl Schell, Chester Evans, Carl Paugh, Mervin Sims, Richard Sherwood and Lee Hartman were the members of the winning team. Sherwood was a digger for the Wolf Den Coal Corporation.

The Brady, Bray and Kimble Coal Company, formed by three former miners, reopened the Wolf Den Mine in 1951, though it was operated at only a minimal level and providing minimal employment for miners. It produced anywhere from 10,000 to 15,000 tons a year, but it was enough to employ a few more miners who had been mostly unemployed for the previous three years.

Jesse and his sister owned this new company and they invested in their company by upgrading the equipment with modern technologies. A 500-foot shaft was dug through Backbone Mountain to allow more air flow through the mine, the ventilating fan was overhauled, mine roads were graded to allow for truck traffic, coal cars were rebuilt and additional land was purchased and leased.

Even with the improvement, though, the miners still used traditional pick mining until 1961 because the coal still needed to be separated from the rock.

One casualty of the upgrades was that the steam power plant at the mine was shut down. Potomac Edison had run a transmission main line from Albright, West Virginia, to Moorefield, West Virginia, that passed over the mining company's property. When the mine reopened, the electric locomotives and arc-cutting machines would be powered by tap from that main line.

Although Jesse Walker was a mine owner, he was also a supporter of union workers because he believed that they benefitted the diggers. However, as the mine's business slowly withered, Jesse had to face the fact that he was either going to have to lay off diggers or reduce their pay. He met with his miners' union rep and laid his cards on the table for them. He told them that he could pay union wages and ship coal. After some back and forth, the men agreed to become nonunion in order to avoid layoffs.

"A little while later, this big Cadillac pulled up with some of the union men from Pittsburgh," Margaret Morris, Jesse's niece, said. "They put a stop to the agreement."

By 1956, only thirty-four homes were left in Shallmar housing 131 residents. Because the houses in Shallmar were constructed from chestnut as some of them were torn down to remove them from the tax rolls, some found second lives as cabins around Deep Creek Lake.

Some of the houses were still company owned, but others were now privately owned by miners who worked somewhere other than Shallmar, but wanted to remain living there or the elderly who didn't need to live near a job. Unlike in 1949, by 1956, all of the homes had radios, twelve had television sets and all of the homes except one had residents with vehicles.

And, yes, miners for BB&K also lived in Shallmar. These faithful miners toiled away in poverty believing that coal demand would pick up again.

Walker and others also believed it would pick up and again.

Walker wanted to be ready. With the improvements he had made, he said that the mine could return to full production within three weeks. However, the mine had not operated at its full production capacity of around 100,000 tons a year since the 1920s.

Frank Powers, the director of the Maryland Bureau of Mines, even bragged to the *Salisbury Times*, that with the additional land that BB&K had leased or bought that the company could maintain full-scale production every day for twenty-five years.

Jesse Walker was getting older as well. He was in a car wreck in the 1960s that slammed his head against the car. The accident caused him to lose his hearing. An examination revealed that he had a benign tumor on his brain. The accident had pressed the tumor against his auditory nerves.

Jesse decided to retire from the mining business and he turned control of the mining property over to the Buffalo Coal Company in 1968. Though he never approved of strip mining, Jesse needed the income when he had surgery to remove the brain tumor.

He lived into his eighties, which was long enough to see the mine shut down for good. Buffalo Coal Company gave mining at Shallmar in 1971. In its fifty-four years of operation, Wolf Den Mine had taken 2,391,945 tons of coal from the Davis six-foot or Lower Kittanning seam and 5,496 tons from Upper Kittanning Seam.

By then, Shallmar had only fifty residents. In 1970, the mountain of coal tailings left behind by the Buffalo Coal Company's mining operations gave way and a mountain of coal and debris slid down the mountain and into Shallmar burying the outhouses and filling the basements of the houses. The coal tailing clogged septic systems in the area and the Maryland State Health Department sent in official to inoculate people against a possible outbreak of typhoid.

Some residents saw this as a sign. "This town's dying anyway. I guess they might as well bury it, too," one woman told a reporter.

20

Santa Leaves Town

Paul left Shallmar at the end of the school year in 1952 when the Garrett County Board of Education closed the school. It was razed a short time later. Paul moved to the West Vindex School for a year and then came to Allegany County in the fall of 1953 as a teacher at Piney Plains School on the eastern side of Allegany County near the border it shares with Washington County. He became principal of the school in 1958.

Piney Plains was another small, country school like Shallmar School. It had three classrooms and three teachers, including Paul who was expected to teach the fifth and sixth graders. Martha was another one of the teachers. She had earned her bachelor's degree in education in 1950 and taught at West Vindex School until 1953. At Piney Plains School, she taught the first and second graders.

Paul stayed at the school as principal until the end of the 1969-1970 school year. He then became principal of Columbia Street Elementary in Cumberland, which was a school at least three times larger than Shallmar School and the largest school where Paul taught.

By the end of his first school year there, Paul could tell that his health was deteriorating. He stepped down as the principal at the school and became a teacher at the Pennsylvania Avenue School, which was a less-stressful position.

However, it didn't allow his healh to improve. He was admitted to Memorial Hospital in Cumberland on December 19 and it was there he died a long way from the town he had helped save.

Paul died on December 28, 1971. He was only fifty five years old. Illness had caused him to spend his last Christmas, the holiday for which he was eternally grateful for the role it played in his life, in the hospital.

Though considered a ghost town as most of the old coal and lumber towns along the North Branch Potomac are called, a few people still live in Shallmar. Of course, most of the houses have been torn down. The company store is still standing, but the owners simply use it for storage.

Robert Hanlin is one of the few residents left. He says he likes the quiet there after having worked in Baltimore for more than forty years. And he gets plenty of quiet except for when the occasional bear comes walking up to his house hoping to find an easy meal in the trash cans.

The only way into Shallmar is still the Shallmar Road, which begins when Main Street in Kitzmiller ends. The road, now paved, runs past a couple dozen of the old miners' houses, the vacant company store building and Howard Marshall's old home. Then it dead ends. A lot of what once was Shallmar is now privately owned by the Wolf Den Hunting Club.

For this community which never had much, there was little to leave behind. The people took with them the most important things – family and the knowledge that no matter how trivial they considered themselves there was a time when the world let each of them know they were loved in a little town called Shallmar.

Shallmar's Santa,
J. Paul Andrick

Notes

In researching this story, there were instances where two or more reputable sources (sometimes even the same source) reported certain details differently. In these instances, I first checked to see if one version was predominant over the other. If so, I generally assumed that to be the correct version and tried to make note of the other version within the notes.

The other thing you will notice is that certain general facts were assumed for Shallmar even if there is no source that specifically notes it. For instance, the Wolf Den Mine was noted in a Bureau of Mines report to be designed in a "room and pillar" style. Another source detailed how "room and pillar" mining was done in a different mine. I applied this method to the Wolf Den Mine since they are both designed the same way.

Chapter 1: A True Story of Santa Claus

p. 1 – *"from nine-month-old ... the oldest man in town"* The ages of these two people were drawn from U.S. Census documents and R. McD, "Garrett Groups Act To Aid Destitute Mining Community." There may have been older or younger people in town at the time, but it's doubtful. Walter Hedrick was fairly close to newborn and with so few families in town, there's a strong chance that he was the youngest resident of Shallmar. Since only Wolf Den Mining Corporation employees lived in Shallmar, it is likely that Howard Marshall was the oldest man in town because anyone older would not have been a working miner, and hence, not a mining company employee.

p. 1 – *"nine if you ... singing about in November"* Santa's most-famous reindeer was first named Rollo, then Reginald, and finally became "Rudolf the Red-Nosed Reindeer." Montgomery Ward Department Stores distributed the story as a booklet around Christmas time in 1939. That first year, 2.4 million copies were distributed. The popularity of the story grew and it became a cartoon and then a song in 1949. Bing Crosby and Dinah Shore turned down the story. Gene Autry finally recorded it and it became the second-bestselling Christmas song behind only Crosby's "White Christmas." Lindquist, Rusty. "The touching story behind Rudolf, the reindeer."

p. 2 – The use of these names for J. Paul Andrick came from the author interviews with Jerry Andrick.

Chapter 2: Towns Come and Go but the Mountain Remains

p. 3 – *"Slaves, criminals and serfs ... it was a punishment."* Lockard, *Coal*, p.9.

p. 3 – *"Country singer Merle Travis ... to the company sto'."* Though Tennessee Ernie Ford's version of "Sixteen Tons" in 1955 is the best known, Merle Travis's version in 1946 was the first. *CowboyLyrics.com*: "Merle Travis: Sixteen Tons."

p. 3 – *"The earliest reference ... are the words 'Coal Mine.'"* Beachley, *History of the Consolidation Coal Co.*, p.2.

p. 3 – *"Farmers mined it part-time ... heat their homes."* Harvey, *The Best-Dressed Miners*, p.5.

p. 4 – *"That all changed ... Cumberland in 1842."* Randolf, *History of the Md. Coal Region*, pp. 514-516.

p. 4 – *"People came from ... at the big event."* Lowdermilk, *History of Cumberland, Md.*, p. 351.

p. 4 – *"The crowd gathered ... Baltimore in ten hours."* *Cumberland Evening Times,* "Baltimore and Ohio Head to Lead Centennial Program."

p. 5 – *"Western Maryland is part ... to be cleaner burning."* Lockard, *Coal*, p. 17.

p. 5 - *"shoot, cut or dig that coal"* Shooting coal referred to using black powder tamped into drilled holes to break up the coal and knock it loose (or shoot it) from the mountain. Cutting coal generally meant the use of machines to cut the coal from the face. Digging coal referred to the method of hand-mining with a pick and shovel. Most mines used a combination of these three techniques.

p. 5 – *"Coal patches tended ... coal seams led."* Lockard, *Coal*, p. 89.

p. 5 – *"Coal companies used ... job in the coal town."* Lockard, *Coal*, p. 89.

p. 6 – *"As one Kempton miner ... had to live accordingly."* Dan Whetzel notes.

p. 7 – *"It started operation in ... April the following year."* Mathews, *Maryland Geological survey*, pp. 272-3.

p. 7 – *"Though an essential commodity ... soldiers fighting in the war."* Savage, *Thunder in the Mountains,* p. x-xi.

p. 7 – *"In fact, the Wolf Den Coal ... in retaliation for the teasing."* George Brady, *Coal Talk.*

p. 7 – *"that wouldn't actually happen ... a woman to be in a coal mine."* President's Commission on Coal, *The American Coal Miner,* p. 191.

p. 7 – *"With the high demand ... wasn't until 1929."* Savage, *Thunder on the Mountain,* p. xi; Maryland Bureau of Mines annual reports.

p. 7 – *"By 1923, U.S. coal ... 142 days that year."* Savage, *Thunder on the Mountain,* p. xi.

p. 7 – *"Shallmar may have started ... and West Virginia, too."* This is something that Brady, McIntyre, Hartman and others mentioned in their interviews.

p. 8 – *"Garrett County, particularly the coal region ... Italians and Poles."* Charles McIntyre, *Dialogs.*

p. 8 – *"The town's population ... wives and children."* This is a rough guess based on the fact that 95 houses were built with an average family size of 5, which also allows for single miners.

p. 9 – *"At 40 years old, Wilbur ... successes to create additional companies."* The list of Marshall's companies came from Hull and Hale, *Coal men of America,* p. 275. Marshall's age came from U.S. Census Data.

p. 9 – *"In 1923, for instance, ... $15,508 a year in today's dollars."* The actual wages came from the *First Annual Report of the Maryland Bureau of Mines,* pp. 20-1. That number was then run through an inflation calculator to find out what the value was in 2012 dollars.

p. 9 – *"Between 1900 ... killed each year in the United States."* Lockard, *Coal,* p. 53.

p. 10 – *"They were only good at face ... they spent their flickers."* Lockard, *Coal,* p. 89.

p. 10 – *"For Saturday's sinners ... of a store and school."* George Brady, *Coal Talk.*

p. 10 – *"a school ... steal from Dodson"* Eary and Grose, *Garrett County Schools of Yesteryear,* p. 487.

p. 10 – *"The town even had rows ... own a car, parking."* Author's George Brady interview.

p. 11 – *"As one coal miner .... coal was ever mined."* Author's George Brady interview.

p. 12 - *"it didn't matter that mine runoff ... wouldn't have cared."*

Maryland Department of the Environment: Report from the Field. An-drick, Hartman and Watts all talked about swimming in red or orange water in the North Branch Potomac and that nothing lived in the water.

p. 12 – *"He sent Italian stonecutters ... torn down just as easily."* Author's Margaret Morris interview.

p. 12 – *"Each summer Wolf Den Coal ... coat of whitewash."* Author interviews with Hartman and Brady.

p. 13 – *"And, it wasn't too long ... directly into their houses."* Both Shirley Watts and Bob Hartman remembered there being running water in their houses in Shallmar, though neither one was sure when this occurred. Newspaper articles at the time note that fact that there were at least some of the houses in Shallmar that were still using the outdoor spigots. Jerry Andrick also noted this in his interview.

p. 13 – *"Coal companies shied away ... could also become expensive."* Sullivan, *Coal and Coal Men,* pp. 173-4.

p. 13 – *"One coal company official once ... than all the rest put together."* Brandt, "Housing the Coal Industry," p. 62.

p. 14 – "So she went around the town ... ripped apart for toilet paper." Using catalogs, especially the Sears and Roebuck catalog for toilet paper declined in the 1930s because the company started printing the catalog on coated, glossy paper. Silverman, *Lindbergh's Artificial Heart,* e-edition.

p. 15 – *"Kenny Bray ... deducted directly from the miner's pay."* From Kenny Bray manuscript.

p. 15 – *"The Shallmar company store ... could be depended on."* From Kenny Bray manuscript.

p. 16 – *"And even on days when ... spelled out 'SHALLMAR.'"* George Brady, *Coal Talk.*

p. 17 - *"Families of Italians ... work in the mines."* Author's Bob Hartman interview; Charles McIntyre, *Dialogs.*

p. 17 – *"Kitzmiller was always a dry town ... brew could put you in a hospital."* From Kenny Bray manuscript.

p. 17 – *"The Wolf Den Mine produced ... Wolf Den Coal Company tenure."* Keller Jr., *Underground Coal Mining in Western Maryland,* p. 364.

p. 17 – *"March 28 dawned ... and into the night."* Kitzmiller High School 1926 Yearbook, p. 87.

p. 18 – *"As residents along the North Branch ... Kitzmiller bridge and became debris."* Kitzmiller High School 1926 Yearbook, pp. 87-8.

p. 18 – "Samuel Beeman, his wife ... tipped over by the floods."

Cumberland Evening Times, "Bodies of Two Flood Victims Are Recovered"; *The News,* "Pittsburgh Dist. Damaged by Flood."

p. 18 – *"The flood began receding ... left of their homes."* Kitzmiller High School 1926 Yearbook, p. 88.

p. 18 – *"Cumberland had $4 million in flood damage."* Feldstein, *Feldstein's Postcard Views,* p. 31.

p. 18 – *"Closer to home, Chaffee ... moved to Vindex."* Garrett County Historical Society, *Ghost Towns of the Upper Potomac,* p. 30

p. 18 – *"In Kitzmiller or Shallmar ... were either destroyed or washed away."* Kitzmiller High School 1926 Yearbook, p. 88.

p. 18 – *"Even after the North Branch ... it hadn't flooded at all."* Garrett County Historical Society, *Ghost Towns of the Upper Potomac,* p. 20

p. 19 – *"Coal had been losing ... electricity became more accepted."* Savage, *Thunder on the Mountain,* p. xi.

p. 19 – *"The second exception to Shallmar's ... for an estimated loss of $28,000."* The Washington Post, "Fire Destroys Tipple of Mine in Maryland."

p. 19 – *"The mine had three ...fresh air each minute."* Mathews, *Maryland Geological survey,* pp. 272-3.

p. 21 – *"Inside, the mine ... experienced and inexperienced miners."* Lockhart, *Coal,* pp. 36-7.

p. 21 – *"That doesn't mean that there weren't ... on December 28, 1932."* Tenth Annual Report of the Maryland Bureau of Mines, p. 20.

p. 21 – *"The Melouse family had ... wrote it down as Melouse."* Author's Margaret Morris interview.

p. 21 – *"While the group of miners ... the Bureau of Mines report read."* Tenth Annual Report of the Maryland Bureau of Mines, p. 20.

p. 21 – *"When an accident happened ... who had been killed."* From Kenny Bray manuscript; Author's George Brady interview.

p. 21 – *"I remember my mother ... It was really sad."* George Brady, *Coal Talk.*

p. 21 – *"Usually the hearse ... deductions made from a miner's pay."* From Kenny Bray manuscript.

p. 22 – *"That was an unpopular ... seen as unpatriotic."* Allegany High School, *Work and Wait,* p. 77.

p. 23 – *"When negotiations failed ... coal continued unimpeded."* Author's George Brady interview.

p. 23 – *"The only change that ... they came to work."* Cumberland Evening Times, "Miners Work Under Ickes-Lewis Truce."

p. 23 – *"Kenny Bray was ... in the future,' Kenny said."* From Kenny Bray manuscript.

p. 24 – *"The armed soldiers ... are wont to be."* Author's George Brady interview.

Chapter 3: A Farmer's Son

Unless otherwise noted, the information about the Andricks comes from the author's interviews with Jerry Andrick.

p. 27 – *"Besides, in those days ... popularity and safety."* Watson, "Teddy Roosevelt: How he saved football, His intervention rescued the sport from its own demise," *History Channel Club*.

p. 27 – *"Since Paul only ... scholarship didn't cover."* Lockard, *Coal*, p. 26.

p. 30 – *"He had registered ... before he was married."* National Archives and Records Administration. *U.S. World War II Army Enlistment Records, 1938-1946.*

p. 31 – *"Camp Blanding was ... added a separation center."* The Camp Blanding Museum and Memorial Park: History.

Chapter 4: Shallmar Declining

p. 32 – *"For instance, telephones ... mudslides and roof collapses."* George Brady, *Coal Talk*.

p. 34 – *"Two coal cars ... coal could be loaded."* From Kenny Bray manuscript.

p. 34 – *"Coal cars weren't designed ... bounce for any riders."* Lockard, *Coal*, p. 4.

p. 34 – *"Mining could be done ... bumping their heads on the ceiling."* Charles McIntyre, *Dialogs*.

p. 35 – *"The diggers usually worked ... which was their only light."* Charles McIntyre, *Dialogs*.

p. 36 – *"The first step ... any other hazards."* Charles McIntyre, *Dialogs*.

p. 36 – *"Elisha Spiker ... killing him instantly."* Fifteenth Annual Report of the Maryland Bureau of Mines, p. 20.

p. 36 – *"Once the timbers were set ... small path open to it."* From Kenny Bray manuscript.

p. 36 – *"Knowing where to drill the hole ... chances of a cave in or rock fall."* Shiflett, *Coal Towns*, p. 87.

p. 37 – *"A squib that acted... coal was checked outside."* From Kenny Bray manuscript.

p. 37 – *"One miner estimated ... 30,000 pounds or 15 tons a day."* Lockard, *Coal*, p. 7.

p. 38 – *"The problem was that explosions ... the pony drivers hauled."* From Kenny Bray manuscript.

p. 39 – *"Charles McIntyre was ... air only inches away."* Charles McIntyre, *Dialogs*.

p. 39 – *"Outside the mine ... know who to yell at."* George Brady, *Coal Talk*.

p. 39 – *"Larger coal operators ... paychecks for dirty loads."* Dan Whetzel notes.

p. 39 – *"Young boys would usually ... main line in Dobson."* George Brady, *Coal Talk*.

p. 40 – *"So the married miners carried ... had our dessert,' he said."* Dan Whetzel notes.

p. 41 – *"They might be missing ... push the coal car."* From Kenny Bray manuscript.

p. 41 – *"George Brady and his future father-in-law... as if he hadn't been injured."* George Brady, *Coal Talk*.

p. 42 – *"A monthly coal charge ... wages for coal"* Author's George Brady interview.

p. 42 – *"The married miners would ... water would get so black from dirt and coal."* Lava soap web site.

p. 42 – *"A bath in those ... too hot or lukewarm."* Author's Shirley Watt interview.

p. 43 – *"Up until this time ... Pittsburgh and West Virginia."* Stegmaier Jr., Dean, Kershaw and Wiseman, *Allegany County: A History*, pp. 283-5, 320-3.

p. 43 – *"Some of the conflicts ... with the pick handle."* From Kenny Bray manuscript.

p. 44 – *"Another time, union miners ... recovering from the beating he took."* Author's George Brady interview.

p. 45 – *"It wouldn't be for ... rest soon followed his lead."* Charles McIntyre, *Dialogs*.

p. 47 – *"Howard had come to Shallmar ... know anything about mining."* Author's Margaret Morris interview.

p. 47 – *"Right before the Wolf Den Coal ... as a "trainman" in New York."* World War I Draft Registration Cards.

p. 47 – *"Even so, he did have to learn ... began requiring in 1923."* *Maryland Bureau of Mines 1923 Annual Report*, p. 93.

p. 47 – *"He was one of the few miners ... dependent on his mining wages alone."* Author's Margaret Morris and George Brady interviews.

p. 47 – *"After a few years as assistant ... bankruptcy process worked itself out."* Maryland Bureau of Mines annual reports.

p. 48 – *"That wasn't such an easy ... company was in receivership."* Garrett County Historical Society, *Ghost Towns of the Upper Potomac*, p. 20; Keller Jr., *Underground Coal Mining in Western Maryland*, pp. 360-5; Maryland Bureau of Mines annual reports.

p. 48 – *"Glosser and Sons Company ... Jesse said when asked."* Author's Margaret Morris interview.

p. 48 – *"For instance, Dottie Crouse ... candy for her kids."* George Brady, *Coal Talk*.

Chapter 5: Shallmar's Principal

Unless otherwise noted, the information about the Andricks comes from the author's interviews with Jerry Andrick.

p. 51 – *"A contemporary of Paul's ... of small-town life."* West Virginia University: WVU Alumni Don Knotts, '48.

p. 51 – *"The most-recent school ... the school to alleviate the overcrowding."* Schools of Yesteryear, p. 331.

p. 51 – *"Paul replaced Hildred Mulvey ... taught himself at the school."* Schools of Yesteryear, p. 333.

p. 52 – *"The Wolf Den Mine was operational ... when the mine shut down."* This information is pieced together from the newspaper articles in the bibliography.

p. 52 – *"Actually, he preferred ... somehow knew if they didn't."* Author's George Brady interview.

Chapter 6: A Kid's Life

Unless otherwise noted, the information about the Andricks comes from the author's interviews with Jerry Andrick.

p. 56 – *"The best deal ... harder to find."* Author's George Brady interview.

p. 57 – *"At the end of a hard day ... apple butter or jelly."* Author's Shirley Watt interview.

p. 57 – *"Though most people enjoy ... from Shallmar stood out."* Author's Bob Hartman interview.

p. 58 – *"The older kids in town ... ignited the powder."* Author's George Brady interview.

p. 61 – *"Some weekends they ... to the day's game."* Author's George Brady interview.

p. 61 – *"During the day ... eyeballed the distance wrong."* Author's Bob Hartman interview.

p. 62 – *"Jesse Walker knew a pair of boys ... to find fish in the river."* Author's Margaret Morris interview.

p. 63 – *"Dottie Crouse would ... wash shattered glasses."* Author's Shirley Watt interview.

Chapter 7: A Man and His Mine

Unless otherwise noted, the information about Jesse Walker comes from the author's interview with Margaret Morris.

p. 64 – *"Jesse was born in Blocton ... eventually have thirteen children."* 1900 U.S. Census.

p. 64 – *"Blocton was a coal mining town ... coal companies in the south for a time."* The Historic Beehive Coke Ovens web site.

p. 64 – *"Shortly after Jesse was born, his family moved further south in Alabama to Wilcox County..."* 1900 U.S. Census.

p. 65 – *"Because of the topography ... little of what went on underground."* Garrett County History: Changing Patterns In Garrett County. *www.deepcreeklake.com/gchs/history/G010911D.htm.*

p. 67 – *"Howard's first marriage ... Jesse became brothers in law."* Charles McIntyre, *Dialogs.*

p. 69 - *"At the beginning of January ... assumed it was one of the laid off miners."* The Washington Post, "Maryland Mine Official Shot in Mystery Attack."

Chapter 8: A Child Faints

p. 70 – *"When he walked over ... United Mine Workers Union Hall."* This comes from both interviews with students and Eary and Grose, *Schools of Yesteryear*, pp. 486-7.

p. 71 – *"Within a few days ... teeter totters for the kids."* Author's Jerry Andrick interviews.

p. 72 – *"As a final touch ... pride in the school."* Author's Jerry Andrick interviews.

p. 73 – *"Mr. Marshall and Jesse ... coal from Maryland Companies."* Cumberland Evening Times, "Coal Purchase By State Can Aid Shallmar" and *Baltimore Sun*, "Foreign Coal Buying Denied."

p. 73 – *"When the Garrett County Commissioners ... operation or very small companies."* Garrett County Commissioner Meeting Minutes, March 8, 1949.

p. 74 – *"The Wolf Den Mine passed its annual inspection ... coal company that didn't have any work."* Oakland Republican, "Wolf Den Mine Inspected Recently."

p. 74 – *"They submitted a bid ... the mine was closed."* Cumberland Evening Times, "Coal Purchase By State Can Aid Shallmar."

p. 75 – *"The town got hit ... there to Kitzmiller School."* Oakland Republican, "Commissioners and School Advisory Board Approve New School Plans."

p. 75 – *"Garrett County had ... Shallmar already attended."* Oakland Republican, "Commissioners and School Advisory Board Approve New School Plans."

p. 77 – The story of the Maule girls comes from a piecing together of stories from the newspaper reports and the author's Jerry Andrick interviews.

Chapter 9: A Town in Need

p. 80 – *"He counted holes ... without shoes and coats."* Author's Jerry Andrick interview.

p. 80 – *"Elva Mae Dean was ... if something was wrong."* The description of Elva Mae Dean comes from looking at pictures of her and author interviews; Eary and Grose, *Schools of Yesteryear*, p. 489.

p. 82 – *"The Wolf Den Coal Company ... afford to pay rent."* From Kenny Bray manuscript.

p. 83 – *"The only sink ... since the mine shut down."* From author interviews.

p. 83 – *"Walter Maule was a ... fewer jobs there would be."* McD., R., "Garrett Groups Act To Aid Destitute Mining Community."

p. 84 – *"The Maules had six children ... killed in a mine accident."* McD., R., "Garrett Groups Act To Aid Destitute Mining Community"; U.S. Census records.

p. 84 – *"The house was well kept .... completely warm in the winter."* McD., R., "Garrett Groups Act To Aid Destitute Mining Community."

p. 85 – *"When mines were operating ... physical labor in the mines."* From Kenny Bray manuscript.

p. 85 – *"Albert Males was ... bankrupt years ago."* Kennedy, "Nation's Gifts Bring Good Will to Hungry Shallmar."

p. 86 – The conversation between Catherine Maule and Paul Andrick is pieced together from newspaper articles which also contained the quotes used in the story.

p. 86 – *"they had been eating only apples ... made them faint."*

This is mentioned in a number of the newspaper articles in the bibliography.

p. 87 – *"The diets of families ... available at all."* Brown and Webb, *Seven Stranded Coal Towns*, pp. 47-8.

Chapter 10: Calling For Help

p. 88 – *"Because the Shallmar miners ... working a three-day week."* The Cumberland Sunday Times, "Short Work Week Gives Miners Glum Outlook as Holiday Approaches." Based on figures in the article, miners would be getting around $46.50 for a three-day work week. The Shallmar miners were getting, at most $25 a week in unemployment benefits, most of them got less.

p. 88 – *"When he had said as much ... to get a new job."* Author interviews with Jerry Andrick and Bob Hartman.

p. 89 – *"The sad thing was ... burned down in the 1920s."* Charles McIntyre, *Dialogs*.

p. 89 – *"The Andrick house ... summer and fall."* Author's Jerry Andrick interview.

p. 89 – *"He walked in the front door ... himself lapse into his thoughts."* Author's Jerry Andrick interview.

p. 90 – *"Then he went into the den ... contracts for the mining company."* Author's Jerry Andrick interview.

p. 91 – *"During the lunch break ... needed a meal."* Author's Shirley Watts interview.

p. 92 – *"It was a large one-story building ... the road was unpaved."* Author's George Brady interview.

p. 92 – *"The company paid miners ... in case there was trouble."* Author's Margaret Morris interview.

p. 92 – *"The building had ... send them on the way."* From the author's George Brady, Jerry Andrick and Bob Hartman interviews.

p. 93 – *"But for miners ... with the mining company."* From Kenny Bray manuscript.

Chapter 11: Finding Her Place

p. 97 – *"During the day ... wringer and left out to dry."* From Kenny Bray manuscript; Charles McIntyre, *Dialogs* and George Brady, *Coal Talk*.

p. 97 – *"Once a week or so ... it tended to go bad quickly."* Compiled from the author interviews.

p. 98 – *"Because the families ... vegetables to her meals."* From

Kenny Bray manuscript.

p. 98 – *"The only money ... severance pay for the miners."* Kennedy, "Nation's Gifts Bring Good Will to Hungry Shallmar."

p. 98 – *"As summer progressed ... ensure that everyone knew."* From Kenny Bray manuscript; Charles McIntyre, *Dialog* and George Brady, *Coal Talk.*

p. 99 – *"There had been an ... I'm doing better than you do."* Author's Jerry Andrick interview.

Chapter 12: On the Brink

p. 100 – *"Each morning, he would take a tally ... if the Andricks had it."* Author's Shirley Watts interview.

p. 101 – *"'Without a great amount of help ... survive the winter,' Paul said."* *Portland Sunday Telegram* and *Sunday Press Herald*, "Shadow of Famine Hangs Over Coal Mining Town."

p. 101 – *"Though Keyser was further ... served two purposes."* Author's Jerry Andrick interviews.

p. 101 – *"The men in town looked ... from December fifth through ninth."* *The Sunday Times*, "Good Hunting Seen For Deer Season In Area."

p. 102 – *"many of Western Maryland's deer ... of targets for hunters."* *The Sunday Times*, "Good Hunting Seen For Deer Season In Area."

p. 102 – *"Hunters from the ... the 1949 season."* Author's George Brady interview.

p. 102 – *"I never cared much for venison ... one woman told a reporter."* *Portland Sunday Telegram* and *Sunday Press Herald*, "Shadow of Famine Hangs Over Coal Mining Town."

p. 102 – *"George Brady bagged ... 100 chickens they had owned."* Kennedy. "Nation's Gifts Bring Good Will to Hungry Shallmar."

p. 103 – *"Not only was Robert Hanlin ... pickled groundhog."* Author's Robert Hanlin interview.

p. 103 – *"Jerry's friend's father, John Crouse ... fed for too long."* Kennedy. "Nation's Gifts Bring Good Will to Hungry Shallmar."

p. 103 – *"He had been a coal miner ... able to support his family."* *The Sun*, "Garrett Groups Act To Aid Destitute Mining Community." and *The American*, "Aid Rushed to Mining Town."

p. 103 – *"It was the first time ... Dolly, said at one point."* *The Sun*, "Garrett Groups Act To Aid Destitute Mining Community."

p. 103 – *"The Crouses were long-time ... midwives in the area."*

The Sun, "Garrett Groups Act To Aid Destitute Mining Community."

p. 103 – *"Mildred Amy Sharpless ... the baby was born."* From Shirley Watts and Bob Hartman interviews.

p. 103 – *"We doctor ourselves ... off what you have to eat."'* Brown and Webb, *Seven Stranded Coal Towns*, p. 46.

p. 104 – *"They had one meal ... and baked beans."* The Sun, "Garrett Groups Act To Aid Destitute Mining Community."

p. 104 – *"Today for a change ... she told one reporter."* Portland *Sunday Telegram and Sunday Press Herald*, "Shadow of Famine Hangs Over Coal Mining Town."

p. 104 – *"The supplemental union help ... stretched that out as well."* Kennedy, "Nation's Gifts Bring Good Will to Hungry Shallmar."

p. 105 – *"Paul had seen students ... had holes in them."* Author's Jerry Andrick interviews.

p. 105 – *"To make matters worse ... from an insulin shock."* Kennedy, "Nation's Gifts Bring Good Will to Hungry Shallmar."

p. 106 – *"Dixie Crosco told a reporter ... was seven years old."* *The American*, "Aid Rushed to Mining Town."

p. 106 – *"I never saw things ... better off than I am."* The American, "Aid Rushed to Mining Town."

p. 107 – *"Their home had only one stove ... that person returned with the lamp."* The Sun, "Garrett Groups Act To Aid Destitute Mining Community."

p. 107 – *"There won't be any Christmas ... lucky to eat."* The American, "Aid Rushed to Mining Town."

Chapter 13: Word Gets Out

p. 108 – *"The Republican had ... all of it."* The Republican Newspaper web site.

p. 109 – *"He had received a letter ... surplus food for the county."* *Oakland Republican*, "Shallmar Residents Are Near Starvation."

p. 109 – *"The Maryland State Police ... he somberly shook his head."* The American, "Aid Rushed to Mining Town"; *Evening Times*, "Aid Offered To Residents Of Shallmar."

p. 109 – *"Briner tried to defend ... very little in 1948."* R. McD., "Mining Town's Outlook Bleak"; *The News*, "Food Being Sent Into Shallmar"; Kennedy, "Nation's Gifts Bring Good Will to Hungry Shallmar."

p. 109 – *"The problem had begun ... twelve in 1949."* The Gettys-

burg Times, "Children In One Mine Town Faint In School From Hunger."

p. 110 – *"Albert Males realized ... Shallmar for their families."* Kennedy, "Nation's Gifts Bring Good Will to Hungry Shallmar."

p. 110 – *"Working on the ladders ... scared these tough miners."* Kennedy, "Nation's Gifts Bring Good Will to Hungry Shallmar."

p. 110 – *"'It would be ... Wolf Den's lack of orders."* R. McD., "Mining Town's Outlook Bleak."

p. 111 – *"The United Mine Workers had called ... $200 a week to $60."* *Cumberland Evening Times*, "Area Miners Prepare For Work Return"; *Cumberland Sunday Times*, "Short Week Givers Miners Glum Outlook As Holiday Approaches."

p. 111 – *"The Shallmar diggers had .... when they had received them."* Kennedy, "Nation's Host of Gifts Bring Good Will to Hungry Shallmar."

p. 111 – *"Most diggers figured that ... considered a bad year."* *Cumberland Evening Times*, "Area Miners Prepare For Work Return"; *Cumberland Sunday Times*, "Short Week Givers Miners Glum Outlook As Holiday Approaches."

p. 111 – *"'It's just no good. ... little left for Christmas.'"* *Cumberland Sunday Times*, "Short Week Givers Miners Glum Outlook As Holiday Approaches."

p. 111 – *"The Oakland American Legion Auxiliary ... handle the distribution."* *Oakland Republican*, "Shallmar Residents Are Near Starvation."

p. 112 – *"Allen Weatherholt ... ran the following day."* Kennedy, "Nation's Gifts Bring Good Will to Hungry Shallmar."

p. 112 – *"At the time, Mr. Marshall ...will be soon.'"* *Cumberland Evening Times*, "Aid Offered To Residents Of Shallmar."

p. 113 – *"Howard was reluctant to talk ... Mining's a gamble."* *Cumberland Evening Times*, "Aid Offered To Residents Of Shallmar."

p. 113 – "At its peak in 1929 ... 100 miners employed." From Maryland Bureau of Mines annual reports.

p. 113 – *"'I ain't seen anyone ... enough taxes,' Mr. Marshall said."* *Cumberland Evening Times*, "Aid Offered To Residents Of Shallmar."

p. 113 – *"While the house ... paid for over a year."* Kennedy, "Nation's Host of Gifts Bring Good Will to Hungry Shallmar."

p. 113 – *"Despite the fact that Mr. Marshall ... separate coal from rock."* *The American*, "Aid Rushed to Mining Town."

Chapter 14: Help

p. 115 – *"Perhaps the biggest surprise ... on Saturday evening."* *The Oakland Republican,* "Food, Clothing Descend Upon Shallmar Miners."

p. 115 – *"Since the ... once they got to town."* *Cumberland Evening Times,* "Combine Your Sunday Drive With Aid For Shallmar People."

p. 116 – *"Seven-year-old Bob Hartman ... it was Christmas morning."* Author's Bob Hartman interview.

p. 117 – *"The group voted to ... would receive larger distributions."* *Cumberland Evening Times,* "Report Shows Much Aid Went To Shallmar."

p. 117 – *"The committee got a ... collections for the area's needy to Shallmar."* *The Oakland Republican,* "Food, Clothing Descend Upon Shallmar Miners."

p. 118 – *"Later in the day ... buy food for Shallmar."* *Cumberland Evening Times,* "Shallmar Appeal For Help Being Answered In District."

p. 118 – *"Two more trucks arrived ... help the residents of Shallmar."* *The Hagerstown Daily Mail,* "Relief Supplies Now Pouring Into Garrett County Coal Mining Town."

p. 118 – *"'I know of one ham ... hid it and so on."* Author's George Brady interview.

p. 118 – *"The Lions and Optimists clubs ... donations that had been collected."* *Cumberland Evening Times,* "Shallmar Appeal For Help Being Answered In District."

p. 118 – *"Paul met with Major Elmer Wall ... sugar and salt pork to the town."* Cumberland Evening Times, "Coal Purchase By State Can Aid Shallmar."

p. 119 – *"Relief efforts for the town ... Paul was accepting donations."* Kennedy, "Nation's Gifts Bring Good Will to Hungry Shallmar."

p. 119 – *"Murrow had been a staple ... lot of people hear it."* Museum of Broadcast Journalism: Murrow, Edward R.

p. 119 – *"George Stonebreaker had ... but now she's off.'"* *The Frederick News,* "54 Shallmar Families Get Aid Taken Into Community."

p. 119 – *"Tony Crosco ... get back to work soon.'"* *The Frederick News,* "54 Shallmar Families Get Aid Taken Into Community."

p. 119 – *"A week after the original ... $500 had been received."*

The Baltimore Sun, "Supplies Grow At Shallmar."

Chapter 15: Playing Political Football
p. 120 – *"That person was Maryland State ... mine was actually in West Virginia."* Cumberland Evening Times, "Coal Purchase By State Can Aid Shallmar."

p. 120 – *"The man who first ... Stanley Coal Company used non-union diggers."* Cumberland Evening Times, "Coal Purchase By State Can Aid Shallmar."

p. 120 – *"Peppered with questions ... operating while other mines shut down."* Cumberland Evening Times, "Coal Purchase By State Can Aid Shallmar;" *The Baltimore Sun,* "'Foreign Coal' Buying Denied."

p. 120 – *"He also pointed out ... in such dire circumstances."* The Baltimore Sun, "'Foreign Coal' Buying Denied."

p. 121 – *"Rennie then explained ... in one basket,' Rennie said."* Cumberland Evening Times, "Coal Purchase By State Can Aid Shallmar"; *The Baltimore Sun,* "'Foreign Coal' Buying Denied."

p. 121 – *"Even with the small contract ... North Branch Potomac once it got started."* Kennedy, "Nation's Host of Gifts Bring Good Will to Hungry Shallmar"; *Cumberland Evening Times,* "Aid Offered To Residents Of Shallmar."

p. 121 – *"He told the reporters, ... diets they need.'"* Cumberland Evening Times, "Coal Purchase By State Can Aid Shallmar."

p. 122 – *"Albert Males told ... Guffey Act in 1935."* Kennedy, "Nation's Host of Gifts Bring Good Will to Hungry Shallmar."

p. 122 – *"Governor Lane had called ... could be taken care of quickly."* The Frederick News, "Kimble To Be On Own With Relief Bills."

p. 122 – *"As the time for the special session ... stretching out for a day or two."* The Frederick News, "Kimble To Be On Own With Relief Bills."

p. 123 – *"On top of this, two delegates ... expense money to the families of Shallmar."* Cumberland Evening Times, "Mine Workers Head Asks Lane To Buy State Coal."

p. 123 – *"John T. Jones ... employees were state residents."* Cumberland Evening Times, "Mine Workers Head Asks Lane To Buy State Coal."

Chapter 16: Children Who Like School Lunches
Unless otherwise noted, most of the information in this chapter

comes from the *Cumberland Times-News* article, "All they knew was that it filled their empty bellies."

p. 125 – *"Then someone ... who used the room as a union hall."* *The Baltimore Sun*, "Garrett Groups Act To Aid Destitute Mining Community."

Chapter 17: Plenty

p. 127 – *"By the middle of December ... to help Shallmar's residents."* *The Baltimore Sun*, "Supplies Grow At Shallmar."

p. 127 – *"'I have answered at least 100 ... near the end of the month."* *The Baltimore Sun*, "Supplies Grow At Shallmar."

p. 127 – *"The Maryland Jewish ... children of Shallmar."* *The Baltimore Sun*, "Supplies Grow At Shallmar."

p. 127 – *"The Amici Corporation ... children of Shallmar."* *The Oakland Republican*, "Sidelights on Aid to Shallmar."

p. 127 – *"Plus, Amici employees raised another $1,000 for the town in general."* *Pittsburgh Press*, "Shallmar Residents Are Near Starvation."

p. 128 – *"The American Legion ... deliver on December seventeenth."* *Cumberland Evening Times*, "State Legion Deluged With Shallmar Aid"; *The Baltimore Sun*, "Gifts Are Ready For Mining Town."

p. 128 – *"From New York City ... things to the town."* *Pittsburgh Press*, "Shallmar Residents Are Near Starvation."

p. 128 – *"The Cumberland ... donating bread to Shallmar."* *Cumberland Evening Times*, "Union, Brewers Help Shallmar."

p. 129 – *"When Paul picked up ... the London Daily Mail."* *Cumberland Evening Times*, "'Shallmar Story' Continues To Tug At Nation's Heart."

p. 129 – *"Letters also arrived ... concern for Shallmar's residents."* Author's Jeanne Hanlin interview.

p. 129 – *"During the distributions ... cry at just the right time."* Author's Jeanne Hanlin interview.

p. 129 – *"About a week after the photo ... has forty—yes, forty— pairs of shoes."* *The Iola Register*, "Aid Pours in to Shallmar, Coal Town Left Destitute by Mine Closure."

p. 129 – *"The picture also spurred ... a new pair of shoes."* *Pittsburgh Press*, "Shallmar Residents Are Near Starvation.

p. 129 – *"Between that free pair ... still better than anything he had."* *Sunday American*, "Air Rushed To Mining Town."

p. 129 – *"The Kitzmiller Boy Scout Troop ... like-new toy on Christmas Day."* The Oakland Republican, "Kitzmiller Troop Activities Lauded At Scout Meeting."

p. 130 – *"On December 21 ... throw the children another party."* Cumberland Evening Times, "'Shallmar Story' Continues To Tug At Nation's Heart."

p. 130 – *"The Republican newspaper ... had been led to believe."* The Oakland Republican, "Sidelights on the Aid to Shallmar."

p. 130 – *"Even some Shallmar ... They're taking charity."* Author's Margaret Morris interview.

p. 130 – *"The Labor Herald... capitalism didn't work."* Allegany Citizen, "Shallmar, John L. Lewis, the UMW & 'Men of Good Will'."

p. 131 – *"The Cumberland Optimist Club ... dozens that was delivering food."* Cumberland Evening Times, "Coal Purchase By State Can Aid Shallmar."

p. 131 – *"Sen. Beall continued ... used for school lunches."* Cumberland Evening Times, "Legion Expands Campaign To Help Counties' Needy."

Chapter 18: A Jolly Christmas

p. 132 – *"The downstairs of the Andrick ... cash and checks in them."* The Iola Register, "Aid Pours in to Shallmar, Coal Town Left Destitute by Mine Closure."

p. 132 – *"On the Friday before ... delivered to the company store."* Cumberland Evening Times, "Aid Continues To Pour Into Garrett Town."

p. 132 – *"The town had received ... other than the relief committee."* Cumberland Evening Times, "Aid Continues To Pour Into Garrett Town."

p. 133 – *"The committee began sending ... and they all got food."* Cumberland Evening Times, "Shallmar Christmas Spirit Spreads to Nearby Hamlets."

p. 133 – *"Today with $5,000 on ... one reporter wrote."* Kennedy, "Nation's Gifts Bring Good Will to Hungry Shallmar."

p. 133 – *"While this was happening ... they wanted."* Kennedy, "Nation's Gifts Bring Good Will to Hungry Shallmar."

p. 133 – *"As each set of parents ... still arriving by truck."* Kennedy, "Nation's Gifts Bring Good Will to Hungry Shallmar."

p. 133 – *"They began ... weren't wild and unruly."* Kennedy, "Nation's Gifts Bring Good Will to Hungry Shallmar."

p. 134 – *"Only five children ... out a 50-cent rag doll."* Kennedy, "Nation's Gifts Bring Good Will to Hungry Shallmar."

p. 134 – *"Shirley Hanlin ... and a new doll."* Author's Shirley Watts interview.

p. 135 – *"Trucks came from ... College View Citizens' Association."* Kennedy, "Nation's Gifts Bring Good Will to Hungry Shallmar."

p. 135 – *"The Montgomery County truck ... couldn't take care of his employees."* Kennedy, "Nation's Gifts Bring Good Will to Hungry Shallmar."

p. 135 – *"That evening, Paul dressed ... with long hair."* Author's Charlotte Chatterton interview.

p. 135 – *"During the Christmas ... from 2 p.m. to 4 p.m."* Cumberland Evening Times, "Aid Continues To Pour Into Garrett Town."

p. 135 – *"'And the person ... events of the day."* Kennedy, "Nation's Gifts Bring Good Will to Hungry Shallmar."

Chapter 19: A Slow Death

p. 136 – *"With hunger only temporarily ... attention to long-term solutions."* Cumberland Evening Times, "Report Shows Much Aid Went To Shallmar."

p. 136 – *"The money, which ... health care, plus other expenses."* Cumberland Evening Times, "Report Shows Much Aid Went To Shallmar."

p. 136 – *"He told the House of Delegates ...Garrett County, though."* The Baltimore Sun, "Shallmar Future Seen In Balance."

p. 137 – *"Leslie Sharpless, the president ... getting much help from the state."* The Republican, "Shallmar Coal Critic Is Blasted."

p. 138 – *"His brother-in-law ... to reopen the mine."* The Oakland Republican, "Retracts Statement on Shallmar Coal."

p. 138 – *"With more than a little fanfare ... did not reopen in 1950."* The Frederick Post, "Shallmar Coal Mine Is Shutdown Once Again."

p. 138 – *"One year to the day ... miners out of work."* The Oakland Republican, "State's Second Largest Mine Plans to Close."

p. 138 – *"A few months later ... 218 miners out of work."* Cumberland Evening Times, State's Largest Coal Mine Shuts Down."

p. 138 – *"The Garrett County Commissioners .... most of them permanently."* Garrett County Commissioners Meeting Minutes, April 24, 1950.

p. 138 – *"Among the many letters ... and so far in 1950 combined."* *Cumberland Evening Times,* "Idle Miners' Families Get Cannery Jobs."

p. 139 – *"Though the work was above ground ... large, he was all right."* Author's George Brady interview.

p. 139 – *"Charlotte Crouse's father ... inexpensive half-house there."* Author's Charlotte Chatterton interview.

p. 139 – *"A ripple effect ... home economics room and a library."* *The Oakland Republican,* "Kitzmiller Area Parents Petition For Improvement To School Plant"

p. 140 – *"A group of parents met ... combined attendance at all four schools."* Gunning, "Garrett School Board Suggests Kitzmiller Group Meet," *Cumberland Evening Times.*

p. 140 – *"This looks like parents ... earn a living for their families."* *The Oakland Republican,* "Of This and That..."

p. 140 – *"Five people were ... certainly aren't going anywhere.'"* Author's George Brady interview.

p. 140 – *"Even the indictments failed ... knew the answer to that question."* *Cumberland Evening Times,* "Indictments Fail To Halt School Strike."

p. 140 – *"When the case came to trial, ... repairs be made to the school."* *Cumberland Evening Times,* "School Strike In Kitzmiller District Ends."

p. 141 – *"At the 1951 Kitzmiller High graduation ... vice principal at Southern High School."* Author's Margaret Morris interview.

p. 142 – *"The following fall fewer than ten ... build four additional classrooms onto the school."* *Cumberland Evening Times,* "Kitzmiller Students Enroll At Elk Garden High School"; *Cumberland Evening Times,* "Rebel School Children Go To Elk Garden."

p. 142 – *"Since the mine at Shallmar ... Fred Baker of the U.S. Bureau of Mines."* From author's George Brady interview; *Cumberland Evening Times,* "Rescue Team Draws Praise From Official."

p. 143 – *"It was the first national championship to be held in twenty years."* *Elyria Chronicle-Telegram,* "400 Enter First Air, Mine Rescue Contests."

p. 143 – *"The competition was held ... equipment, physical condition and ability."* *Newark Advocate and American Tribune,* "Mine Rescue Teams Go Into Action Tonight."

p. 143 – *"At the end of the competition... digger for the Wolf Den Coal Corporation."* *Cumberland Evening Times,* "Rescue Team Draws

Praise From Official."

p. 144 – *"The Brady, Bray and Kimble ... separated from the rock."* Keller Jr., *Underground Coal Mining in Western Maryland,* p. 364; *The Salisbury Times,* "Maryland Town Continues Its Hard Struggle For Existence."

p. 144 – *"By 1956, there were only ... housing 131 residents"* The *Salisbury Times,* "Maryland Town Continues Its Hard Struggle For Existence."

p. 144 – *"Because the houses in Shallmar ... around Deep Creek Lake."* Author's Margaret Morris interview.

p. 145 – *"Even Howard Marshall had ... residents with vehicles."* *The Salisbury Times,* "Maryland Town Continues Its Hard Struggle For Existence."

p. 145 – *"Frank Powers, the director ... even they gave up on it in 1971."* *The Salisbury Times,* "Maryland Town Continues Its Hard Struggle For Existence."

p. 145 – *"Jesse Walker was getting older ... mine shut down for good."* Author's Margaret Morris interview.

p. 145 – *"In its fifty-four years ... from Upper Kittanning Seam."* Keller Jr., *Underground Coal Mining in Western Maryland,* p. 365.

p. 145 – *"By then, Shallmar had only ... one woman told a reporter."* Jay, "County Caught Between Coal, Tourism," *The Washington Post.*

Chapter 20: Santa Leaves Town

p. 146 – *"Paul left Shallmar ... another one of the teachers."* The *Cumberland News,* "Former Local Principal Dies In Hospital."

p. 146 – *"She had earned her bachelor's degree ... West Vindex School until 1953."* Eary and Grose, *Garrett County Schools of Yesteryear,* p. 500.

p. 146 – *"At Piney Plains ... first and second graders."* Cumberland Evening Times, "Andrick New School Head.

p. 146 – *"Paul died on December ... from the town he had helped save."* The Cumberland News, "Former Local Principal Dies In Hospital."

Bibliography

Newspapers

Allegany Citizen. (Md.) Unsigned article. "Shallmar, John L. Lewis, the UMW & 'Men of Good Will'." December 22, 1949.

The Baltimore News-Post. (Md.) Mullikin, James C. "U.S. Aid Sought To Save Shallmar." December 12, 1949.

The Baltimore Sun. (Md.) Ferguson, Ernest B. "Life Struggles On In Ghost Towns." July 15, 1956.

----------. McD., R. "Mining Town's Outlook Bleak." December 10, 1949.

----------. "Garrett Groups Act To Aid Destitute Mining Community." December 11, 1949.

----------. "'Foreign Coal' Buying Denied." December 14, 1949.

----------. "Supplies Grow At Shallmar." December 16, 1949.

----------. "Shallmar Families Get Donated Food, Clothing." December 14, 1949.

----------. "Gifts Are Ready For Mining Town." December 17, 1949.

----------. "3 Trucks Of Legion-Gathered Contributions Go To Shallmar." December 18, 1949.

----------. "St. Nick Relieves Shallmar's Plight." December 25, 1949.

----------. "Shallmar Future Seen In Balance." February 15, 1950.

The Baltimore Sunday American. (Md.) "Aid Rushed to Mining Town." December 10, 1949.

The Burlington Daily Times-News. (N.C.) Unsigned article. "Generous Public Aids Destitute Mining Town." December 15, 1949.

Cumberland Evening Times. (Md.) Gunning, Eugene T. "Garrett School Board Suggests Kitzmiller Group Meet." September 8, 1950.

---------- Unsigned article. "Bodies of Two Flood Victims Are Recovered." April 4, 1924.

----------. Unsigned article. "Baltimore and Ohio Head to Lead Centennial Program." November 5, 1942.

----------. Unsigned article. "Miners Work Under Ickes-Lewis Truce." May 4, 1943.

----------. Unsigned article. "Kitzmiller." January 5, 1944.

----------. Unsigned article. "Miners' Appeal on Jobless Pay is Heard Here." April 17, 1946.

----------. Unsigned article. "Area Miners Prepare For Work Return." December 1, 1949.

----------. Unsigned article. "New Records Established For Deer Kills." December 7, 1949.

----------. Unsigned article. "Garrett Town Facing Dark Holiday Period." December 9, 1949.

----------. Unsigned article. "Aid Offered To Residents of Shallmar." December 10, 1949.

----------. Unsigned article. "Shallmar Appeal For Help Being Answered In District." December 11, 1949.

----------. Unsigned article. "Legion Expands Campaign To Help Counties' Needy." December 12, 1949.

----------. Unsigned article. "Help Arrives In Quantity For Shallmar." December 12, 1949.

----------. Unsigned article. "Coal Purchase By State Can Aid Shallmar." December 13, 1949.

----------. Unsigned article. "Kitzmiller Troop Activities Lauded At Scout Meeting." December 14, 1949.

----------. Unsigned article. "Life Stirs Again In Shallmar But Men Still Lack Mine Work." December 14, 1949. Jack F. Davis

----------. Unsigned article. "Mine Workers Head Asks Lane To Buy State Coal." December 14, 1949.

----------. Unsigned article. "'Shallmar Story' Continues To Tug At Nation's Heart." December 15, 1949.

----------. Unsigned article. "State Legion Deluged With Shallmar Aid." December 16, 1949.

----------. Unsigned article. "Unions, Brewers Help Shallmar." December 20, 1949.

----------. Unsigned article. "Shallmar Christmas Spirit Spreads to Nearby Hamlets." December 22, 1949.

----------. Unsigned article. "Aid Continues To Pour Into Garrett Town." December 25, 1949.

----------. Unsigned article. "Shallmar Men Urged To Find Other Work." February 15, 1950.

----------. Unsigned article. "Kimble To Ask Lane About Coal." February 15, 1950.

----------. Unsigned article. "Rennie Blasted For Statement On

Shallmar Coal." February 16, 1950.

----------. Unsigned article. "Over Half Of States' Coal Mines Closed." April 5, 1950.

----------. Unsigned article. "Idle Miners' Families Get Cannery Jobs." June 16, 1950.

----------. Unsigned article. "Hardesty Announces Faculty For Garrett County Schools." August 29, 1950.

----------. Unsigned article. "Indictments Fail To Halt School Strike." September 13, 1950.

----------. Unsigned article. "Boycott Holds At Kitzmiller." September 18, 1950.

----------. Unsigned article. "School Strike In Kitzmiller District Ends." September 27, 1950.

----------. Unsigned article. "Students Back As Kitzmiller Strike Ends." September 27, 1950.

----------. Unsigned article. "Rescue Team Draws Praise From Official." October 5, 1951.

----------. Unsigned article. "Kitzmiller Lions Fete Mine Drill Team Champs." October 22, 1951.

----------. Unsigned article. "Report Shows Much Aid Went To Shallmar." June 26, 1952.

----------. Unsigned article. "Kitzmiller Students Enroll At Elk Garden High School." September 3, 1952.

----------. Unsigned article. "Rebel School Children Go To Elk Garden." September 4, 1952.

----------. Unsigned article. "Rescue Team In Action." October 13, 1954.

----------. Unsigned article. "Andrick New School Head." August 8, 1958.

Cumberland News. (Md.) Unsigned article. "Former Local Principal Dies in Hospital." December 29, 1971.

----------. Unsigned article. "Myron H. McDonald." May 20, 1972.

The Cumberland Sunday Times. (Md.) Bolton, Jim. "Short Week Gives Miners Glum Outlook As Holiday Approaches. " December 4, 1949.

----------. Unsigned article. "Shallmar." May 23, 1943.

----------. Unsigned article. "Good Hunting Seen For Deer Season In Area." December 4, 1949.

----------. Unsigned article. "Shallmar Appeal For Help Being Answered In District." December 11, 1949.

----------. Unsigned article. "Local And Tri-State Obituaries," May 29, 1955.

Cumberland Times-News. (Md.) Rada Jr., James. "When Coal was King." September 5, 2002.

----------. Rada Jr., James. "All they knew was that it filled their empty bellies." August 30, 2004.

The Elyria Chronicle-Telegram.(Ohio) "400 Enter Mine Rescue, First Air Contests." October 2, 1951.

The Frederick News. (Md.) Unsigned article. "Exceptions to Distribution." October 7, 1943.

----------. Unsigned article. "Food Being Sent Into Shallmar." December 12, 1949.

----------. Unsigned article. "54 Shallmar Families Get Aid Taken Into Community." December 14, 1949.

----------. Unsigned article. "Kimble To Be On Own With Relief Bills." December 14, 1949.

----------. Unsigned article. "Shallmar Food Pile Growing." December 16, 1949.

----------. Unsigned article. "Shallmar Residents Must Find Other Work Or Stay On Relief." February 15, 1950.

----------. Unsigned article. "To Re-Open Shallmar Mine When Strike Ends." February 16, 1950.

----------. Unsigned article. "Shallmar Defended By Garrett Senator." February 17, 1950.

----------. Unsigned article. "Western Maryland Coal Towns Are 'Just So Many Shallmars'." April 5, 1950.

The Frederick Post. (Md.) Unsigned article. "Response Amazing." December 16, 1949.

----------. Unsigned article. "American Legion Given Supplies For Shallmar." December 17, 1949.

----------. Unsigned article. "People From Every State But 1 Answered Shallmar's Appeal." January 11, 1950.

----------. Unsigned article. "Shallmar Mine Will Be Reopened After Strike." February 17, 1950.

----------. Unsigned article. "Shallmar Coal Mine Is Shutdown Once Again." April 5, 1950.

----------. Unsigned article. "Western Maryland Coal Towns Are 'Just So Many Shallmars'," April 6, 1950.

The Gettysburg Times. (Pa.) Unsigned article. "Children In One Mine Town Faint In School From Hunger." December 13, 1949.

James Rada, Jr.

Grant County Press. (Petersburg, W. Va.) Unsigned article. "Shallmar Showered With $20,000 Worth of Christmas Gifts." January 18, 1950.

The Hagerstown Daily Mail. (Md.) Unsigned article. "Relief Supplies Now Pouring Into Garrett County Coal Mining Town." December 12, 1949.

The Hagerstown Morning Herald. (Md.) Unsigned article. "Food And Clothing Pour Into Shallmar," December 4, 1949.

The Huntingdon News. (Pa.) Unsigned article. "Pittsburgh Dist. Damaged by Flood." March 31, 1924.

The Iola Register. (Kan.) Unsigned article. "Aid Pours in to Shallmar, Coal Town Left Destitute by Mine Closure." *January 10, 1950.*

Lewiston Daily Sun (Maine) Unsigned article. "Happy Holidays Assured Shallmar." December 24, 1949.

Los Angeles Times. (Calif.) Meyer, Eugene A. "Old Mining Town, Likes Its Widows, Clings to Life." November 28, 1979.

The Massillon Evening Independent. (Ohio) "Hold Rescue Contest At Scene of Mine 'Disaster.'" October 2, 1951.

The Mineral Daily News Tribune. (Keyser, W. Va.) Unsigned article. "Kitzmiller Miner Carried To Death By Conveyor Belt." December 21, 1949.

New York Times. (N.Y.) Unsigned article. "Wilbur H. Marshall." January 4, 1967.

Newark Advocate and American Tribune. (Ohio) "Mine Rescue Teams Go Into Action Tonight." October 2, 1951.

Oakland Republican (Md.) Unsigned article. "Wolf Den Mine Inspected Recently," May 5, 1949.

----------. Unsigned article. "Rumbling of Trouble Around Local Mines." September 29, 1949.

----------. Unsigned article. "Board of Education Makes Annual Report." November 24, 1949.

----------. Unsigned article. "Commissioners And School Advisory Board Approve New School Plans." December 1, 1949.

----------. Unsigned article. "Shallmar Residents Are Near Starvation." December 8, 1949.

----------. Unsigned article. "366 Deer Killed First Three Days of Six-Day Season." December 8, 1949.

----------. Unsigned article. Special Session of Legislature Called." December 8, 1949.

----------. Unsigned article. "Shallmar's Plight." December 15, 1949.

----------. Unsigned article. "Food, Clothing Descend Upon Shallmar Miners." December 15, 1949.

----------. Unsigned article. "Sidelights on Aid to Shallmar." December 15, 1949.

----------. Unsigned article. "Money, Food and Clothes Continue To Reach Mining Town." December 22, 1949.

----------. Unsigned article. "Young Miner Is Killed Near Vindex." December 22, 1949.

----------. Unsigned article. "Writer Interested In Shallmar Story." January 26, 1950.

----------. Unsigned article. "May Discontinue Main Mail Train." February 9, 1950.

----------. Unsigned article. "Shallmar Coal Critic Is Blasted." February 16, 1950.

----------. Unsigned article. "Retracts Statement on Shallmar Coal." February 23, 1950.

----------. Unsigned article. "Mines Operating." March 9, 1950.

----------. Unsigned article. "State's Second Largest Mine Plans To Close." March 16, 1950.

----------. Unsigned article. "Kempton Mine Closes After 38 Year's Operation." April 20, 1950.

----------. Unsigned article. "Of This and That..." June 8, 1950.

----------. Unsigned article. "Kitzmiller Area Parents Petition For Improvement To School Plant." June 8, 1950.

Pittsburgh Press. (Pa.) Beachler, Edwin. "Hearts, Purses Open, Save Starving Village After Mine Closes." December 17, 1949.

----------. Unsigned article. "Shallmar Residents Are Near Starvation." December 18, 1949.

Portland Sunday Telegram and Sunday Press Herald. (Maine) Unsigned article. "Shadow of Famine Hangs Over Coal Mining Town." December 11, 1949.

Poughkeepsie New Yorker. (N.Y.) Unsigned article. "Shallmar Shares Its Bounty." December 22, 1949.

Reno Evening Gazette (Nev.) Unsigned article. "Jobs Still Lacking For Mining Town," December 15, 1949.

The Salisbury Times (Md.) Unsigned article. "Maryland Town Continues Its Hard Struggle For Existence." October 29, 1956.

Spokane Daily Chronicle (Wash.) Unsigned article. "Life Stirring Again in Little Coal Town as Food Arrives." December 14, 1949.

The St. Joseph Herald-Press. (Mich.) Unsigned article. "Shallmar

Gets Lots of Help." January 10, 1950.

The Stars and Stripes. Unsigned article. "Supplies Sent to Aid Town Impoverished By Closed Mine." December 12, 1949.

Tucson Daily Citizen. (Ariz.) Unsigned article. "Shallmar Residents Are Told Destitution To Be Permanent." February 16, 1950.

The Washington Post. (D.C.) Jay, Peter A. "County Caught Between Coal, Tourism." April 12, 197.

----------. Meyer, Eugene L. "Old Towns Clings to Life." October 28, 1979.

----------. Richards, Bill. "Coal Mining Suit May Cost State Millions in Claims." November 9, 1973.

----------. Unsigned article. "Fire Destroys Tipple of Mine in Maryland." November 13, 1930.

----------. Unsigned article. "Maryland Mine Official Shot in Mystery Attack" January 11, 1933

----------. Unsigned article. "Free-Staters Pour Food Into Shallmar." December 17, 1949.

----------. Unsigned article. "Ghost Town Relief Fund Books Closed." June 28, 1952.

The Washington Star. (D.C.) Kennedy, George. "Nation's Gifts Bring Good Will to Hungry Shallmar." December 24, 1949.

----------. Unsigned article. "Nation's Gifts Bring Good Will to Hungry Shallmar." December 24, 1949.

Wisconsin Rapids Daily Tribune. (Wis.) Unsigned article. "Shallmar Has Life Again; One Thing Lacking—Work." December 14, 1949.

Youngstown Vindicator. (Ohio) Unsigned article. "Voice of New-Born Christ Calls to World, Brings Out Finest Things in Man This Day." December 25, 1949.

Interviews and Unpublished Manuscripts

Andrick, Jerry. Interviewed by the author in person and via e-mail beginning on September 9, 2011.

Brady, George. *Coal Talk: Dialogs with people from Western Maryland Coal Communities.* Interviewed by Gail Herman on June 8, 1991. For Oral History Workshop Sponsored by Garrett Recreation, Parks and Tourism with partial funding from Garrett County Development Corporation and Maryland Humanities Council with a grant from the National Endowment of the Humanities.

Brady, George. Interviewed by the author on January 29, 2012.

Bray, Kenny. An unpublished manuscript about life in a mining town written by Kenny Bray who was a Shallmar miner in the 1940s to 1960s.

Chatterton, Charlotte. Interviewed by the author on January 29, 2012.

Hanlin, Jeanne. Interviewed by the author on January 29, 2012.

Hanlin, Robert. Interviewed by the author on November 8, 2011.

Hartman, Robert. Interviewed by the author on January 29, 2012.

McIntyre, Charles. *Dialogs with People from Western Maryland Coal Communities*, Interviewed by Andrea Hammer on Oct. 3, 1991.

Morris, Margaret. Interviewed by the author on March 23, 2012.

Watts, Shirley. Interviewed by the author on January 29, 2012.

Whetzel, Dan. Notes he made to a draft of this book. April 5, 2012.

Books

Allegany High School Oral History Class. *Work and Wait: Allegany County: The Home Front Years 1941-1945.* (Cumberland, Md.: Commercial Press, 2003).

Beachley, Charles E. Beachley. *History of the Consolidation Coal Co., 1864-1934.* (New York, New York: The Consolidation Coal Company, 1934).

Brown, Malcolm and Webb, John N. *Seven Stranded Coal Towns: A Study of an American Depressed Area.* (New York, N.Y.: Da Capo Press, 1971).

Eary, Alice Ann Feather and Grose, Jean Taylor Williams. *Garrett County Schools of Yesteryear.* (Oakland, Md. : Garrett County Historical Society, 2008).

Feldstein, Albert. *Feldstein's Top Historic Postcard Views of Allegany County, Maryland.* (Cumberland, Md.: Commercial Press Printing Company, 1997).

Garrett County Historical Society (compiled by) *Ghost Towns of the Upper Potomac.* (Parsons, W.Va.: McClain Printing Company, 1998).

Giesen, Carol A. B. Coal Miners' Wives: Portraits of Endurance. (Lexington, Kent.: The University Press of Kentucky, 1995).

Harvey, Katherine A. *The Best-Dressed Miners: Life and Labor in the Maryland Coal Region, 1835-1910.* (Ithaca, N.Y.: Cornell University Press, 1969).

Hull, Arthur M. and Hale, Sydney A. *Coal Men of America: A Biographical and Historical Review of the World's Greatest Industry.* (Chicago, Ill.: The Retail Coalman, 1918).

Keller Jr., Vagel Charles. *Underground Coal Mining in Western Maryland, 1876-1977: A Reference Guide.* (Frostburg, Md.: Lewis J. Ort Library, Frostburg State University, 2008).

Kitzmiller High School. *Kitzmiller High School 1926 Yearbook.* (Kitzmiller, Md., 1926).

Lockard, Duane. *Coal: A Memoir and Critique.* (Charlottesville, Va.: University Press of Virginia, 1998).

Lowdermilk, William Harrison. *History of Cumberland, Md.* (Washington, D. C.: James Anglim, 1878).

Randolph, Beverly S. *History of the Md. Coal Region.* V, 514-6. (Baltimore, Md.: Maryland Geological Survey, 1905).

Savage, Lon. *Thunder in the Mountains: The West Virginia Mine War 1920-21. (Pittsburgh, Pa.: The University of Pittsburgh Press, 1990).*

Schlosnagle, Stephen and the Garrett County Bicentennial Committee. *Garrett County A History of Maryland's Tableland.* 2nd edition. (Parsons, W.Va.: McClain Printing Company, 1997).

Silverman, Steve. *Lindbergh's Artificial Heart: More Fascinating True Stories from Einstein's Refrigerator.* (Kansas City, Mo.: Andrews McMeel Publishing, 2003) e-book edition.

Stegmaier Jr., Harry; Dean, David; Kershaw, Gordon; and Wiseman, John. *Allegany County: A History.* (Parsons, W. Va.: McClain Printing Company, 1976).

Sullivan, Charles Kenneth. *Coal Men and Coal Towns: Development of the Smokeless Coalfields of Southern West Virginia, 1873-1923.* (New York, N.Y.: Garland Publishing, 1989).

Magazines and Journals

Frank, E. R. "John L. Lewis and Roosevelt's Labor Policy." *Fourth International*, Vol.4, No.4, April 1943, pp.102-106. *www.marxists-fr.org/history/etol/writers/cochran/1943/04/lewis.htm.*

Watson, Bruce. "Teddy Roosevelt: How he saved football, His intervention rescued the sport from its own demise." *History Channel Club.* *www.thehistorychannelclub.com/articles/articletype/articleview/articleid/1429/teddy-roosevelt-how-he-saved-football.*

Reports and Documents

Ancestry.com. *1880 United States Federal Census* [database on-line].

Provo, UT: Ancestry.com Operations, Inc. 1999. Original data: United States of America, Bureau of the Census. Tenth Census of the United States, 1880. Washington, D.C. : National Archives and Records Administration, 1880. T9, 1,454 rolls.

----------. *1900 United States Federal Census* [database on-line]. Provo, UT, USA: Ancestry.com Operations Inc., 2004. Original data: United States of America, Bureau of the Census. *Twelfth Census of the United States, 1900*. Washington, D.C.: National Archives and Records Administration, 1900. T623, 1854 rolls.

----------. *1910 United States Federal Census* [database on-line]. Provo, UT, USA: Ancestry.com Operations Inc., 2006. Original data: *Thirteenth Census of the United States, 1910*. Washington, D.C.: National Archives and Records Administration, 1910. T624, 1,178 rolls.

----------. *1920 United States Federal Census* [database on-line]. Provo, UT, USA: Ancestry.com Operations Inc., 2010. Original data: *Fourteenth Census of the United States, 1920*. Washington, D.C.: National Archives and Records Administration, 1920. T625, 2076 rolls.

----------. *1930 United States Federal Census* [database on-line]. Provo, UT: Ancestry.com Operations Inc., 2002. Original data: United States of America, Bureau of the Census. *Fifteenth Census of the United States, 1930*. Washington, D.C.: National Archives and Records Administration, 1930. T626, 2,667 rolls.

Brandt, L. "Housing the Coal Industry." *Proceedings*. (West Virginia Coal Mining Institute, 27th and 28th Annual Sessions, 1923).

Garrett County Commissioners. 1949 Meeting Minutes. Garrett County Commissioners Offices. Oakland, Md.

Maryland Board of Labor and Statistics. *First Annual Report of the Maryland Bureau of Mines of the State of Maryland, Covering Period May 1, 1922, to December 31, 1922, and Period January 1, 1923, to December 31, 1923*. (Baltimore, Md.: 20th Century Printing Company, 1924).

----------. *Second Annual Report of the Maryland Bureau of Mines of the State of Maryland, Calendar Year 1924*. (Baltimore, Md.: 20th Century Printing Company, 1925).

----------. *Third Annual Report of the Maryland Bureau of Mines of the State of Maryland, Calendar Year 1925*. (Baltimore, Md.: 20th Century Printing Company, 1926).

----------. *Fourth Annual Report of the Maryland Bureau of Mines of the State of Maryland, Calendar Year 1926.* (Baltimore, Md.: 20th Century Printing Company, 1927).

----------. *Fifth Annual Report of the Maryland Bureau of Mines of the State of Maryland, Calendar Year 1927.* (Baltimore, Md.: 20th Century Printing Company, 1928).

----------. *Seventh Annual Report of the Maryland Bureau of Mines of the State of Maryland, Calendar Year 1929.* (Baltimore, Md.: 20th Century Printing Company, 1930).

----------. *Eighth Annual Report of the Maryland Bureau of Mines of the State of Maryland, Calendar Year 1930.* (Baltimore, Md.: 20th Century Printing Company, 1931).

----------. *Ninth Annual Report of the Maryland Bureau of Mines of the State of Maryland, Calendar Year 1931.* (Baltimore, Md.: 20th Century Printing Company, 1932).

----------. *Tenth Annual Report of the Maryland Bureau of Mines of the State of Maryland, Calendar Year 1932.* (Baltimore, Md.: Press of King Bros., 1933).

----------. *Twelfth Annual Report of the Maryland Bureau of Mines of the State of Maryland, Calendar Year 1934.* (Baltimore, Md.: 20th Century Printing Company, 1935).

----------. *Thirteenth Annual Report of the Maryland Bureau of Mines of the State of Maryland, Calendar Year 1935.* (Baltimore, Md., 1936).

----------. *Fourteenth Annual Report of the Maryland Bureau of Mines of the State of Maryland, Calendar Year 1936.* (Baltimore, Md., 1937).

----------. *Fifteenth Annual Report of the Maryland Bureau of Mines of the State of Maryland, Calendar Year 1937.* (Baltimore, Md., 1938).

----------. *Sixteenth Annual Report of the Maryland Bureau of Mines of the State of Maryland, Calendar Year 1938.* (Baltimore, Md., 1939).

----------. *Seventeenth Annual Report of the Maryland Bureau of Mines of the State of Maryland, Calendar Year 1939.* (Baltimore, Md., 1940).

----------. *Eighteenth Annual Report of the Maryland Bureau of Mines of the State of Maryland, Calendar Year 1940.* (Baltimore, Md., 1941).

----------. *Nineteenth Annual Report of the Maryland Bureau of Mines of the State of Maryland, Calendar Year 1941.* (Baltimore, Md.,

1942).

----------. *Twentieth Annual Report of the Maryland Bureau of Mines of the State of Maryland, Calendar Year 1942.* (Baltimore, Md., 1943).

Maryland Board of Natural Resources, Department of Geology, Mines and Water. Resources, *Thirty-First Annual Report of the Maryland Bureau of Mines of the State of Maryland, Calendar Year 1953.* (Baltimore, Md., 1954).

Mathews, Edward Bennett. *Maryland Geological Survey, Vol. 11* (Baltimore, Md.: Johns Hopkins University Press, 1922).

National Archives and Records Administration. *U.S. World War II Army Enlistment Records, 1938-1946* [database on-line]. Provo, UT, USA: *Ancestry.com* Operations Inc., 2005. Original data: *Electronic Army Serial Number Merged File, 1938-1946* [Archival Database]; World War II Army Enlistment Records; Records of the National Archives and Records Administration, Record Group 64; National Archives at College Park, College Park, MD.

President's Commission on Coal, John D. Rockefeller IV, chairman. *The American Coal Miner: A Report on Community and Living Conditions.* (Washington D.C., 1980).

Source Watch: Maryland and coal. *www.sourcewatch.org/index.php?title=Maryland_and_coal.*

United States, Selective Service System. World War I Selective Service System Draft Registration Cards, 1917-1918. Washington, D.C.: National Archives and Records Administration. M1509, 4,582 rolls. Imaged from Family History Library microfilm.

Web Sites

The Camp Blanding Museum and Memorial Park: History. *www.campblanding-museum.org/History/History.htm.*

CowboyLyrics.com. "Merle Travis: Sixteen Tons." *www.cowboylyrics.com/tabs/travis-merle/sixteen-tons-1369.html.*

Critical Past: Mine workers listen to news over a radio about coal mines being taken over by the government in the United States. *www.criticalpast.com/video/65675060481_mine-workers_news-broadcast-over-radio_copy-of-President%e2%80%99s-order_flashlight-helmets.*

Garner, Mike. Maryland Department of the Environment: Report from the Field: Shallmar Coal Refuse Reclamation Project.

www.mde.state.md.us/programs/ResearchCenter/ReportsandPubl ica-tions/Pages/ResearchCenter/publications/general/emde/vol2no1/s hallmar.aspx.

Garrett County History: Changing Patterns In Garrett County. *www.deepcreeklake.com/gchs/history/G010911D.htm.*

The Historic Beehive Coke Ovens, Blocton, AL. *www.wbbm.org/history-west-blocton.htm.*

Internacional Jose Guillermo Carrillo Foundation Library of Voices of the Twentieth Century: Franklin D. Roosevelt Speeches, On the Coal Crisis Fireside Chat, May 2, 1943. *www.fundacionjoseguillermocarrillo.com/sitio/documentos/FRD/ On%20the%20Coal%20Crisis,%20(May%202,%201943).pdf.*

Lava Soap FAQs, *www.lavasoap.com/faq.*

Lindquist, Rusty. "The touching story behind Rudolf, the reindeer." *Life Engineering, life-engineering.com/1958/the-touching-story-behind-rudolf-the-reindeer.*

Maryland Department of the Environment. Maryland Abandoned Mine Cleanup Efforts Win National Award. *www.mde.state.md.us/programs/PressRoom/Pages/936.aspx*

Murrow, Edward R. Museum of Broadcast Communications. *www.museum.tv/eotvsection.php?entrycode=murrowedwar.*

The Republican Newspaper: Who we are. *www.therepublicannews.com/videotour.aspx.*

Western Maryland's Historical Library. *www.whilbr.org.*

West Virginia University. "WVU Alumni Don Knotts, '48." *alumni.wvu.edu/awards/academy/don_knotts*

About the Author

James Rada, Jr. is the author of seven novels, a non-fiction book and a non-fiction collection. These include the historical novels *Canawlers, October Mourning, Between Rail and River* and *The Rain Man.* His other novels are *Logan's Fire, Beast* and *My Little Angel.* His non-fiction books are *Battlefield Angels: The Daughters of Charity Work as Civil War Nurses, Looking Back: True Stories of Mountain Maryland* and *Looking Back II: More True Stories of Mountain Maryland.*

He lives in Gettysburg, Pa., where he works as a freelance writer. Jim has received numerous awards from the Maryland-Delaware-DC Press Association, Associated Press, Maryland State Teachers Association and Community Newspapers Holdings, Inc. for his newspaper writing.

If you would like to be kept up to date on new books being published by James or ask him questions, he can be reached by e-mail at *jimrada@yahoo.com.*

To see James' other books or to order copies on-line, go to *www.jamesrada.com.*

If you liked
SAVING SHALLMAR,
you can find more stories at these FREE sites from James Rada, Jr.

JAMES RADA, JR.'S WEB SITE
www.jamesrada.com

The official web site for James Rada, Jr.'s books and news including a complete catalog of all his books (including eBooks) with ordering links. You'll also find free history articles, news and special offers.

TIME WILL TELL
historyarchive.wordpress.com

Read history articles by James Rada, Jr. plus other history news, pictures and trivia.

WHISPERS IN THE WIND
jimrada.wordpress.com

Discover more about the writing life and keep up to date on news about James Rada, Jr.

Made in the USA
Columbia, SC
24 September 2019